ENHANCED OIL RECOVERY

Field Planning and Development Strategies

VLADIMIR ALVARADO

EDUARDO MANRIQUE

AMSTERDAM • BOSTON • HEIDELBERG • LONDON • NEW YORK • OXFORD
PARIS • SAN DIEGO • SAN FRANCISCO • SINGAPORE • SYDNEY • TOKYO

Gulf Professional Publishing is an imprint of Elsevier

Gulf Professional Publishing is an imprint of Elsevier
30 Corporate Drive, Suite 400
Burlington, MA 01803, USA

The Boulevard, Langford Lane
Kidlington, Oxford, OX5 1GB, UK

Notices

Knowledge and best practice in this field are constantly changing. As new research and experience broaden our
understanding, changes in research methods, professional practices, or medical treatment may become
necessary.
   Practitioners and researchers must always rely on their own experience and knowledge in evaluating and
using any information, methods, compounds, or experiments described herein. In using such information or
methods they should be mindful of their own safety and the safety of others, including parties for whom they
have a professional responsibility.
   To the fullest extent of the law, neither the Publisher nor the authors, contributors, or editors, assume any
liability for any injury and/or damage to persons or property as a matter of products liability, negligence or
otherwise, or from any use or operation of any methods, products, instructions, or ideas contained in the
material herein.

Library of Congress Cataloging-in-Publication Data
Alvarado, V. (Vladimir)
   Enhanced oil recovery : field planning and development strategies —
V. Alvarado and E. Manrique
      p. cm.
   Includes bibliographical references and index.
   ISBN 978-1-85617-855-6
   1. Enhanced oil recovery. I. Manrique, E. (Eduardo) II. Title
TN871.37.A485 2010
622′.33827–dc22   2010012466

British Library Cataloguing-in-Publication Data
A catalogue record for this book is available from the British Library.

For information on all Gulf Professional Publishing publications
visit our Web site at *www.elsevierdirect.com*

Printed in the United States
11 12 13 14   10 9 8 7 6 5 4 3 2

To our loving and patient wives,
Teresa and Elimar

Contents

7. Methodology   113

8. EOR's Current Status   133

9. Closing Remarks   157

Preface

Developing a book of any magnitude requires time away from family and friends, so as we finish this one, we must confess that despite the obvious intellectual and personal satisfactions we enjoyed while writing this book, it was not always a pleasant experience. The book's contents are for the most part the result of scribbling on napkins over numerous macchiati and espressos away from the office at different posts over the years. As researchers and consultants, perhaps our most creative moments arose during lengthy informal, and somewhat dreamy, discussions about enhanced oil recovery. That is why many projects came to fruition after consuming many heavily caffeinated cups of coffee.

We have taken a practical approach to describing our thoughts on decision making when applied to enhanced oil recovery (EOR). We know that EOR requires patience, perseverance, and (yes, we admit it) stubbornness, but the final goal is field implementation. Our modest contribution to decision making is aimed at facilitating and encouraging more EOR activities.

ACKNOWLEDGMENTS

We are indebted to numerous colleagues for their contributions in the form of ideas, encouragement, support, and friendship. Aaron Ranson and his team were a creative force behind the efforts to develop screening technologies that simultaneously accommodate both objectivity and practicality—not at all a simple demand. Several staff members of the Oil Recovery Methods Department at a former company provided the necessary feedback as we struggled to find solutions to EOR decision problems. The dedication of young and senior engineers and geologists to many of the EOR projects we participated in generated some of the input for our analyses. We would like to acknowledge some of those colleagues and friends.

Tamara Liscano patiently looked at numerous databases, making sure everything made as much sense as possible. A number of colleagues offered critiques (some that were not always gentle) of our efforts, to which Guillermo Calderon and José Manuel Alvarez can probably relate. E.-M. Reïch, K. Potsch, Y. Yunfeng, and L. Surguchev generously shared their thoughts for a number of years. Jane and John Wright provided a nurturing atmosphere at Questa Engineering, where many fields were evaluated, and improvements to our methods soon followed.

Two Questa junior colleagues and collaborators, Mehdi Izadi and Curtis Kitchen, patiently generated modeling data and tested some of our most recent ideas. Our joint article served as the starting point for this book, and we will always be thankful for their efforts. Many thanks go to Mahdi Kazempour, a graduate student at the University of Wyoming, for providing simulation data and plots.

We are truly indebted to our editor Ken McCombs and to Elsevier for the opportunity to publish this book. It was certainly a matter of serendipity, but no doubt Ken found value in some of our ideas.

Last, but not least, our families have been supportive and patient to the extreme. Teresa knows what this means to Vladimir, and she has always worked to make a home wherever the family has moved. Elimar, Anjuli, and Eduardo Andres are certainly proud of Eduardo, just as he is of them.

We probably have forgotten to mention many colleagues and friends who were sources of inspiration and ideas. We know they will forgive us for this, understanding that they are always in our thoughts.

Introduction

This book explains strategies for evaluating reservoir development plans (RDPs) based on enhanced oil recovery (EOR). In this sense, it focuses on the decision-making that leads to launching EOR projects. In the context of this book, any strategy that ultimately increases oil and gas recovery is under consideration for EOR decisions. The definitions of EOR will be explored in detail, but the authors introduce important concepts through examples and by briefly reviewing the evolution and history of these methods. Although only a modest fraction of global oil production (3 to 5 percent) can be attributed to EOR, a number of oil provinces in the world rely on it as the main recovery mechanism. This trend will very likely see an increase running apace with a decrease in the number of discoveries and the sizes of hydrocarbon pools, or as new discoveries are made in harsher environments such as deepwater offshore locations.

We examine both already completed and ongoing reported projects to exemplify the value of proper decision making in EOR. The authors have been working in the oil and gas industry in several upstream segments, including research and development and planning and execution of pilot projects, as well as in support activities as consultants for major oil companies and small operators for more than 20 years. A resource, and central theme, here is the workflow that came to light after many years of professional practice, which resulted from the need to develop tools and procedures to deal with improved EOR decision making.

The oil market in recent years has triggered a significant increase in property evaluation and acquisition and development of enhanced oil recovery projects. This upsurge in EOR activities has been motivated not only by an invigorated oil market, which remains relatively strong despite an economic slowdown, but also by, to a great extent, better-known provinces reaching maturity and the possibility of increasing reserves in well-known locations.

In perspective, out of the 3 trillion barrels of oil known to exist in conventional reservoirs, only one-third have been produced and consumed in the market since the early times of the oil business. An additional one-third of the oil in place is expected to be produced by techniques beyond traditional oil and gas activities, including advanced, but commercially viable, EOR. Entire conferences and conference sessions have been dedicated to this issue in recent times, and it is likely to become even more relevant at future meetings; the 2009 SPE Research and Development Conference is a good example.

Future sustainable hydrocarbon production will involve combining yields from both unconventional resources and fields in harsher environments such as deep offshore and politically and/or ecologically sensitive areas. Digital technologies have been predicted to become a large part of the any solution related to the next-trillion problem (Miller, 2008; Moon, 2008). These technologies include automation, data mining, and smart-field technologies.

One important consideration while producing this book was the scarcity of properly trained personnel who can deal with some of the decision challenges associated with EOR. The lack of required teams of engineers and geoscientists can be associated with the oil price collapse during the 1980s and with the later phase-out of R&D centers in major oil companies. There are only a few groups at well-recognized universities and oil companies that continue to develop, evaluate, and/or understand the key features of EOR technologies today.

This state of affairs in our industry has strongly impacted EOR decision making over the last two decades, leading to delays and, probably, missed opportunities when it comes to increasing oil recovery. The main factor impacting financial investments in EOR operations is oil price volatility. EOR initiatives are often delayed under these conditions because of either perceived or real financial risk.

Time is also an issue for EOR decision making. If you are unfamiliar with EOR recovery mechanisms and the known consequences of delaying implementation decisions, it is important for you to develop a sense about the window of opportunity. For example, a common naive conclusion, usually resulting from incorrectly framed financial decisions, is to postpone EOR projects until the economic limit of primary or secondary projects has been attained. This type of decision making assumes that favorable conditions for EOR activities found in a given reservoir at a given time will prevail for the rest of reservoir's productive life. Another way of looking at this is by considering analyses that lead to decisions. For instance, it is a good idea to use a variety of screening methods as part of your decision-making framework. If screening is executed once and never reviewed as reservoirs evolve, you might be left with scenarios with expiration dates.

To exemplify the window of opportunity, or the time issue, consider a screening exercise for a miscible process. (*Miscibility* refers to the ability of two or more fluids to mix at the molecular level.) For example, your can of soda is bubbly because carbon dioxide (CO_2) is dissolved at high pressure in the liquid. But as soon as the can is popped open, the CO_2 comes out, and eventually the soda goes flat. This process is very similar to the loss of energy that occurs in reservoirs as the pressure is depleted and the oil becomes "flat," with frequent undesirable consequences.

Now, let us go back to the issue of decisions in the face of time. In essence, the ability of a solvent (e.g., carbon dioxide) to efficiently sweep the oil-containing pores it contacts very much depends on pressure in the case of gases. As we will see, a condition, variable, or parameter that impacts reservoir recovery this way is referred to as *critical*. The fact that miscibility is so important for recovery means, in practice, that pressure is a critical variable. If the reservoir pressure remains higher than the so-called minimum miscibility pressure (i.e., the value of the pressure required for dissolving the solvent in the oil phase), can the injection fluid (solvent) be a good recovery agent? If the reservoir's initial pressure is adequate for a miscible process, then a screening exercise will likely show it to be a good candidate for this recovery strategy.

Such screening procedures should not be used to produce a go or no-go answer but should provide a feasibility determination on the basis of only a few relevant rock and fluid variables, typically the critical ones. Now, for instance, if a viable miscible EOR process at time t is delayed because it is simply less expensive to produce under primary or secondary recovery strategies (i.e., for purely economic reasons), the window of opportunity for miscible EOR might be missed, even if it was originally technically viable. This is a consequence of the reservoir's energy (i.e., pressure) being depleted irreversibly for lack of pressure-maintenance mechanisms.

As a result, reservoirs do not remain static during any exploitation phase, and so the time allotted for a decision in EOR is constrained. This is not as uncommon as you might think. To help you to understand the underlying problems, the revision of reservoir development plans are discussed in Chapter 1.

Another case is property acquisition, which involves limited time for making a decision. Overanalyzing a purchase without introducing new, relevant data, however, can destroy the value of an acquisition because the chances for success can diminish if the number of decisions is perilously insufficient (Begg, Bratvold, and Campbell, 2003).

Most likely, one of the reasons that overanalysis has become so deeply rooted in the oil and gas industry is that analysis through detailed modeling can reduce uncertainty. The belief that numerically accurate reservoir dynamic models can overcome the hurdles of ambiguity, or even uncertain data sources, is groundless. Modeling should be the least complex as possible to support rational decision making. Bos (2005) shows that lower precision and a higher level of modeling of uncertainty and integration might be necessary to optimize the E&P decision-making process. This may be attainable at the expense of a trade-off between the degree of "model precision" and the degree of uncertainty modeling and integration in favor of the latter.

The oil and gas industry presents its own peculiarities with respect to decision making (Mackie and Welsh, 2006). A pressing issue in decision-making problems is framing (Skinner, 1999), which helps to lower ambiguity with respect to goals or even to eliminate conflicting objectives by developing a decision hierarchy, strategy tables, and an influence diagram (see Chapter 6). In practice, framing signifies knowing exactly what the focus of the decision is and, just as important, what it is not. The importance of understanding the EOR decision focus cannot be overstated, so it is crucial that the object of EOR decision-making exercises be clearly defined to avoid a fishing expedition.

One of the difficulties with decision making is risk avoidance, which is as much a trait of humans as it is a characteristic of organizations. As the complexity of field operations increases, risk avoidance in decision makers triggers "the overanalysis loop." When this occurs, decision makers resort to increasing levels of analysis and modeling or simulation in the hope that uncertainty will be reduced and the possibility of undesirable outcomes can be lowered to negligible levels. The mistake with this view is that uncertainty is not the same as ambiguity, so ill-defined objectives are often confused with lack of certainty. If critical data are not available, analysis will not provide the desired certainty. Even when the decision-making process is rational and reasonable, the outcome can still be negative.

Pedersen, Hanssen, and Aasheim (2006) discuss qualitative screening and soft issues, which are important considerations in EOR analysis and decision making. Petroleum and, more specifically, reservoir engineering professionals focus on the quantitative analysis of production mechanisms and on the evaluation of reserves and performance (reservoir simulation), among many other analytical tasks. Decision making relies on the quantifiable aspects of a problem, such as the net present value of the project, so rational decisions can be made. The difficulty arises when unquantifiable issues become part of the decision problem.

Social and environmental considerations often present themselves as qualitative aspects of a problem, which can be difficult to put into quantifiable terms. For EOR, sources of raw materials (e.g., water), disposal of by-products or waste, and proximity to sinks and sources frequently barely become quantifying matters and must be incorporated into the analysis as soft issues. Retraining of analysts is then necessary to weigh in some of these considerations so that resources are not unnecessarily committed to hard analysis before barriers associated with soft issues are overcome or at least understood.

Ensuring that the model focuses on relevant decision criteria is a prerequisite for overall model relevance. The point is that NPV or other economic (hard) indicators should be used for hard, quantifiable issues, while a variety of methods can be implemented to address soft issues. In this way, the balance between the two provides a good basis for decision alternatives. A balanced analysis of soft and hard issues is an important aspect of decision making discussed in this book.

The oil and gas industry devotes much effort to complex analyses of uncertainty quantification, hoping to eliminate, or at least reduce, it. Bickel and Bratvold (2007) present the results of a survey of decision makers, support teams, and academics to define the value of uncertainty quantification in decision making. The Society of Petroleum Engineers (SPE) as a professional community has held a significant number of forums on uncertainty evaluation but few on decision making. This might explain why such an intense focus has been place on uncertainty analysis as a goal in itself.

One conclusion from Bickel and Bratvold's survey is that the complexity of decision analysis has not greatly contributed to improving the decision-making process in our industry, at least as perceived by those who responded to the survey. The decision-analysis cycle can also be considered iterative in the sense that if more assessments are required (or if profitable data are being gathered), then the information should be compiled and the cycle repeated.

Another frequently encountered problem in decision making is the use of "expert opinion." That the answer came from an expert on the subject, does not necessarily make it correct. Often, excessive use of intuition, which can be mistaken for expertise, can create significant bias. Although intuition may very well have its place in decision making (Dinnie, Fletcher, and Finch, 2002), it can hurt the decision-making process itself. For example, the chemical flooding problem in the 1970s caused many to declare that the processes being used were not sufficient for the commercial market.

Despite the technical merits attributed to the designs produced by the research laboratories involved, they were deemed economic failures. Today, new chemistry and process designs have produced a significant number of technical and economical successes for chemical flooding operations. Thus, the ability to determine what is necessary to make chemical flooding both economically feasible and technically viable has improved considerably.

An additional important consideration in EOR decision making is cognitive bias (Welsh, Bratvold, and Begg, 2005). This can take many forms, one of which reflects the cognitive limitations of the human mind (Begg, Bratvold, and Campbell, 2003). The level of risk avoidance may not be consistent with goals, objectives, and prudent decision making.

This is patently clear when value is destroyed because the decision maker's aversion to risk is higher than the organization's.

A number of methodological strategies have been developed over the years to deal with decision making for EOR projects. In Goodyear and Gregory's studies (1994), screening based on critical variables for the enhanced oil recovery processes is used to determine feasibility early on in an evaluation.[1] This step, however, should not be performed before the problem is framed, including some important soft issues (e.g., local availability of resources or even experience in EOR deployment). EOR decision making must be considered a continuous exercise in screening and scoping (preliminary economics) to provide the best combination of soft and hard issues as inputs for decision makers. In this sense, it is often found that data gathering is one of the most recommended courses of action.

To mitigate cognitive bias, several different database approaches are needed. Data-mining strategies can be used as part of advanced screening with this intent in mind. Thus, instead of relying on a few experts' biases, numerous biases are incorporated into the framed decision problem as emerging from the data structure. EOR screening techniques have been widely documented in the literature. Most of them rely on conventional and advanced approaches (Al-Bahar et al., 2004; Guerillot, 1988; Henson, Todd, and Corbett, 2002; Ibatullin et al., 2002; Joseph et al., 1996). However, very few studies focus on the decision-making process initiated from well-documented screening exercises.

This book provides elements of decision making that are tailored to EOR practices to give readers and practitioners the tools necessary to become more effective at deploying EOR projects. Elements of successful enhanced oil recovery methods and fundamental concepts are discussed to serve as background materials for readers who are unfamiliar with modern EOR technologies. The steps making up a flexible screening methodology are included, as well as details on various analytical and numerical simulation approaches that can be used for different field studies as part of the continuous development of the proposed EOR screening methodology. Performance estimations by means of simplified models illustrate a wide range of decision opportunities, as highlighted by Bos (2005). The case studies are based on examples from the authors' research and consulting practice.

Chapter 1 reviews reservoir development plans as the starting point for EOR decisions. Chapter 2 provides some important definitions associated with EOR and oil recovery concepts. Chapter 3 discusses the elements of reservoir simulation, most of which focus on analytical

---

[1] The decision-making workflow that is discussed in this book was partially inspired by Goodyear and Gregory's work.

simulation. Chapter 4 examines screening methods for EOR, which are a central aspect of the methodology for decision making. Chapter 5 presents important decision criteria based on soft issues. Chapter 6 provides elements of framing and discusses the tools used for this purpose and the fundamentals of financial evaluation.

Chapter 7, which is this book's pivotal chapter, describes the workflow used for EOR decision making. If you have not read the earlier chapters and are unfamiliar with these topics, we suggest you scan them. Chapter 8 reviews the current status of enhanced oil recovery in general. It is a practical summary that should help you integrate the ideas in the book and understand future EOR goals. Numerous references—some of which are not cited in this book—are provided in the last section. We hope that readers will find that the list adds extra value to the important subject of enhanced oil recovery.

Reservoir Development Plans

Numerous publications have been dedicated to reservoir development planning and integrated reservoir management (Babadagli et al., 2008; Bibars and Hanafy, 2004; Cosentino, 2001; Dudfield, 1988; Fabel et al., 1999; Figueiredo et al., 2007; Gael et al., 1995; Satter and Thakur, 1994; Schiozer and Mezzomo, 2003; Stripe et al., 1993). This book provides a general overview of reservoir development planning to set the context for evaluating and implementing enhanced oil recovery (EOR) projects. In other words, reservoir development planning refers to strategies that begin with the exploration and appraisal well phase and end with the abandonment phase of a particular field to establish the course of action during the productive life of the asset. Figure 1.1 summarizes the phases of a reservoir development plan. The main objective of the complete cycle of a development plan is to maximize the asset value.

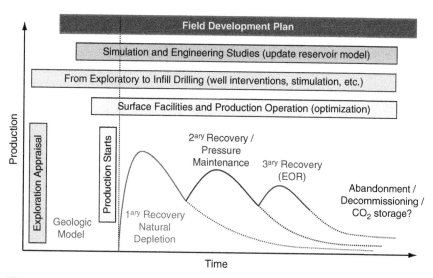

FIGURE 1.1   The main phases of a field development plan.

Enhanced Oil Recovery
DOI: 10.1016/B978-1-85617-855-6.00007-3

Development strategies for new fields are based on data obtained from seismic surveys (which are not always acquired or readily accessible), exploratory wells, and other limited information sources such as fluid properties and reservoir analogues. Based on the information at hand, initial development plans are defined through simulation studies considering either a probabilistic or a stochastic approach to rank options using economic indicators, availability of injection fluids (i.e., water and/or gas), and oil recovery and risk, among other considerations.

Therefore, integrating the information from simulation studies helps to address the multiple and complex factors that influence oil recovery, as well as reservoir development decisions. As new information about the reservoir, its geology, and its degree of heterogeneity becomes available through drilling of new wells (i.e., development and infill wells) and production–injection history, the field can be developed in an optimal way.

In the case of mature fields with a steady decline in oil production, new development plans must be reevaluated or implemented. However, if the decision to implement a new development plan in mature fields is made too late (i.e., fields producing with oil cuts below 5 percent), the number of economically viable options becomes limited. This case relates to the value of time or the window of opportunity for implementing EOR projects in mature fields.

For a variety of reasons, most, if not all, reservoir development plans (RDPs) change or must be adjusted or modified during the productive life of the field. Some of the reasons include the following:

- Lack of reservoir characterization and understanding of production mechanisms at the early stages of development (reduction of uncertainties with time)
- Poor production performance (e.g., production below expectations and early water breakthrough)
- Environmental constraints or drivers (e.g., CO_2 storage, changes in legislation)
- Economics (e.g., low oil prices)
- New technologies (e.g., horizontal wells, multilaterals, and new recovery processes)

Thus, dynamic and flexible reservoir management is required to optimize field production responses that maximize the value of the asset over its full cycle of exploitation.

Considering again the importance of time and reservoir pressure in development plans, Figure 1.2 presents a simple decision tree to evaluate the potential applicability of different recovery processes in light to medium crude oil reservoirs. (We will discuss influence diagrams and decision trees in the Chapter 6, Economic Considerations and Framing.)

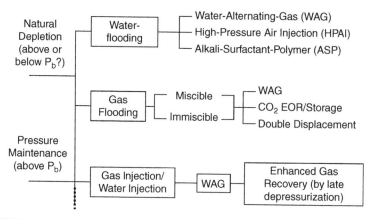

FIGURE 1.2   A simplified example of a decision tree to evaluate the potential recovery processes as part of the RDP in light to medium crude oil reservoirs.

Although Figure 1.2 does not show steam injection methods, although it is still a valid recovery process (Perez-Perez et al., 2001).

In general, a particular light or medium oil reservoir can be a suitable candidate for several EOR processes, as we will see later in the book. However, if pressure maintenance (either by water or gas injection) starts below the bubble point pressure (P_b), the probability of obtaining lower ultimate oil recovery increases compared to the case of reservoirs in which the secondary recovery initiates at pressures above P_b. Additionally, timing for pressure maintenance as part of a reservoir development plan can be critical to control variables such as the following:

- Asphaltene deposition/flocculation because of their impact on reservoir performance, well injectivity, and/or well productivity (Civan, 2007; Garcia et al., 2001; Kabir and Jamaluddin, 2002; Poncet et al., 2002).
- Retrograde condensation, which is typical of gas and condensate reservoirs when pressure goes below the dew point (Belaifa et al., 2003; Briones et al., 2002; Clark and Ludolph, 2003).
- Problems with sand production and wellbore collapse and stability (Bellarby, 2009; Civan, 2007; Nouri et al., 2003; Tovar et al., 1999).

Enhanced oil recovery chemical methods such as alkali-surfactant-polymer (ASP) have gained considerable interest in recent years as these methods have matured and become commercial options to increasing oil recovery in mature waterfloods. To demonstrate the impact of past decisions on the future technical and, most important, economic success of chemical EOR processes, Figure 1.3 shows an example of some of the decisions an operator generally faces when planning a water injection project in a particular field.

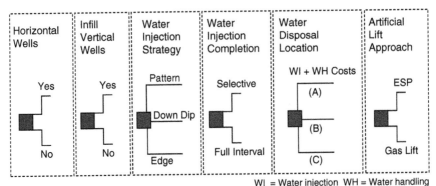

WI = Water injection WH = Water handling

FIGURE 1.3   A simplified decision analysis for a waterflooding project.

Specifically, in recent project evaluations the authors have completed, well spacing has been one of the biggest hurdles of economic feasibility of chemical EOR processes. In some project evaluations, we have found that infill drilling programs are needed or recommended to accelerate oil recovery and thus the rate of return on investment.

We will touch on this type of strategy in association with the larger issue of improved oil recovery, or IOR. However, incremental oil recovery that is estimated during EOR chemical flooding project evaluations is not always sufficient to pay off capital expenditures associated with drilling programs, reducing the upside potential of mature waterflooded reservoirs. The latter combined with the volatility of crude oil prices represents a big challenge in RDPs of mature fields.

On the other hand, reservoir development plans for heavy and extra-heavy crude oil reservoirs, including oil sands, generally differ from those of medium and light crude oil reservoirs. Given the viscosity of heavy oils at reservoir conditions, oil might not flow naturally. This is the case of Canadian oil sands and some tar sands in other areas of the world (viscosities on the order of 10^6 cp). In oil sands, EOR technologies such as steam-assisted gravity drainage, or SAGD, are necessary to produce oil sands at economic rates. In these cases, EOR can be used earlier in the sequence of reservoir development plans of heavy and extra-heavy oils. Thus, EOR methods should not always be associated with tertiary recovery methods as shown in Figure 1.1.

Figure 1.4 shows the elements of a simple decision tree with some of the options of recovery processes that are potentially applicable in heavy to extra-heavy crude oil reservoirs. This particular example is based on the flow of viscous oil at reservoir conditions. It is not surprising that EOR thermal methods represent the most common recovery processes envisioned and applied to develop heavy and extra-heavy oil reservoirs.

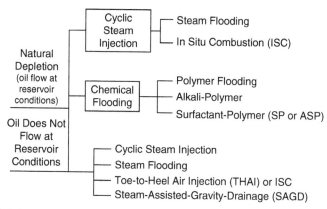

FIGURE 1.4 A simplified example of a decision tree to evaluate potential recovery processes as part of the RDP in heavy to extra-heavy (oil sands) crude oil reservoirs.

However, several pilot tests of chemical EOR processes applied to heavy-oil reservoirs—that is, ASP and alkali-polymer (AP)—have been documented in the literature in recent years. These tests have opened a new window of opportunity for heavy crude oil ($14° < API < 22°$) reservoirs (Arihara et al., 1999; Pitts et al., 2004; Pitts et al., 2006; Pratap and Gauma, 2004; Zhijian et al., 1998).

As you may have realized by now, reservoir development decisions create a history for a reservoir that has a significant impact on decisions down the road for EOR opportunities. We have indicated through examples that a tertiary application of EOR technologies is not a must, and it turns out that the earlier you deploy EOR, the better if the objective is total optimum recovery in terms of volumes of hydrocarbon.

We elaborate on enhanced oil recovery definitions and mechanisms in this book to highlight their impact on decision making, and we would like to demystify some of those "expert opinions." Despite limited production associated with EOR processes in most productive areas, these technologies are out of the lab and are currently being applied in numerous fields. It is just a myth that EOR represents an opportunity only for the distant future. Enhanced oil recovery not only provides a way to increase reserves, which is loosely defined as oil you can extract by commercial means, but it also might offer an economic way to prolong the productive life of assets and delay the decommissioning stage that most companies abhor.

2

Enhanced Oil Recovery Concepts

2.1 INTRODUCTION

In this chapter, we examine some of the principles of enhanced oil recovery (EOR) methods and highlight some of the most important market and technical drivers for EOR. Before we can discuss decision making in EOR, we must be certain that we understand exactly what EOR is. Many different schools of thought come into play when discussing EOR decision making. Legislation, convenience, and other factors often determine the appropriate course of action, but ingenuity and flexibility are also important considerations. Classification offers only a framework rather than the means to an end.

This chapter deals with the current concept of recovery mechanisms, which we connect to their relative groups. Instead of providing a purely academic approach to this, we discuss these concepts as groups of methods. Good books on the fundamentals of porous media and reservoir engineering dedicate more space to some of these topics than we can practically do in this book. We believe, however, that context is everything in EOR, so you will have to establish the context of your specific situation and then analyze the decision-making process within that context.

Enhanced Oil Recovery
DOI: 10.1016/B978-1-85617-855-6.00008-5

We take the opportunity here to elaborate on the issue of timing and deployment of EOR methods. From traditional definitions of enhanced oil recovery, it is suggested that these strategies should be initiated after the primary and the secondary methods' economic performance have been exhausted. As we have already mentioned, if certain reservoir conditions are exceeded, certain viable processes in a reservoir at the early stages of production might not be technically and economically feasible later on. We will use examples to elaborate on this process and as we discuss the workflow for EOR decision making.

2.2 WHAT IS ENHANCED OIL RECOVERY?

We have been talking about EOR-related topics with only a broad understanding of exactly what enhanced oil recovery is. If any concept is a source of heated controversy, it is EOR. This book does not distinguish between traditional approaches to EOR and other technologies associated with the concept of improved oil recovery (IOR) for a good reason. Now, you might argue that we are evading a definition of EOR by introducing a new term, which is partially correct; however, as you will see, there is a need to treat these two ideas concurrently. EOR and IOR often intertwine, and as a result, new and more effective ways to improve recovery spring from the initial attempts to establish or deploy either route. As you will see, IOR encompasses EOR, and this creates a superset of strategies and technologies for oil production that are superior to traditional methods— namely, water flooding and gas flooding.

You should also keep in mind that IOR/EOR has several drivers. From the point of view of reservoir engineers and others, gains in reserves (or recovery factors) and/or productivity can guide decisions in both EOR and IOR. On the other hand, decision makers can be motivated by legal or financial reasons. It is not uncommon to receive tax incentives to launch EOR initiatives by federal or state entities, which makes it convenient to define the operation as an EOR process. Examples of this arose from attempts to develop screening of EOR processes in Wyoming (Alvarado et al., 2008).

A particular set of reservoir–field combinations (i.e., those associated with the Minnelusa formation) were analyzed to determine the best EOR options for these reservoirs. If you access the oil and gas databases in Wyoming, you will discover that polymer injection, which is traditionally associated with EOR, was in fact used as a well conformance agent (e.g., gels for water control). Examples of the data sources include the following:

- The Wyoming Geological Association (WGA) field guidebooks and symposiums (Wyoming Geological Association Publications, 1946–2000)
- The Wyoming Oil and Gas Conservation Commission (WOGCC) production database (The Wyoming Oil and Gas Commission, 2008a and 2008b)
- The database of crude oil analyses performed by the National Institute for Petroleum and Energy Research (NIPER) and now hosted by the Department of Energy's National Energy Technology Laboratory (DOE/NETL) (National Energy Technology Laboratory, 1983)
- The National Petroleum Council Public Database, also hosted by the DOE/NETL (National Energy Technology Laboratory, 1984)
- The Rocky Mountain Basins Produced Water database, compiled and hosted by the DOE/NETL (National Energy Technology Laboratory, 2005)

All of the processes filed by operators were labeled as EOR, including the necessary waterflooding phase. Most likely this was done for tax-break reasons. So it is not that the operators never intended to perform EOR, but that the correct drivers may have led to increased activity in this area, and a flexible definition might be beneficial. However, when legal constraints clearly define an EOR process, those guidelines must be obeyed, or an exemption must be filed.

2.2.1 IOR and EOR Definitions

Let us establish the definitions of IOR and EOR so that we have a clear idea of what we are dealing with in this book. What is here is in no way meant as comprehensive coverage of the literature, but instead provides selected published definitions that will allow us to build consistency and flexibility. We analyze what follows for the sake of clarity.

The Society of Petroleum Engineers, or SPE (SPE E&P Glossary, 2009) offers the following definitions:

1. *Improved oil recovery*, or IOR, is "any of various methods, chiefly reservoir drive mechanisms and enhanced recover(y) techniques, designed to improve the flow of hydrocarbons from the reservoir to the wellbore or to recover more oil after the primary and secondary methods (water- and gasfloods) are uneconomic."
2. *Primary oil recovery* is "the amount of the reserves recovered by primary production—that is, without injected fluid pressure support."
3. *Secondary oil recovery* is "a recovery improvement process such as waterflooding or gasflooding."
4. *Enhanced oil recovery*, or EOR, is "one or more of a variety of processes that seek to improve the recovery of hydrocarbon from a reservoir after the primary production phase."

Carcoana (1992) describes the intent of EOR methods as follows:

1. To improve sweep efficiency by reducing the mobility ratio between injected and in-place fluids.
2. To eliminate or reduce capillary and interfacial forces and thus improve displacement efficiency.
3. To act on both phenomena simultaneously.

Satter and colleagues (2008) define EOR as

... relat[ing] to advanced processes to further augment oil recovery beyond secondary recovery (by waterflooding or natural gas injection) in a reservoir. Enhanced oil recovery processes include all methods that use external sources of energy and/or materials to recover oil that cannot be produced economically by conventional means.

Thomas (2008) refers to both IOR and EOR as follows:

1. Improved oil recovery, or IOR, "is a general term which implies improving oil recovery by any means."
2. Enhanced oil recovery, or EOR, "implies a reduction in oil saturation below residual oil saturation (Sor)."

The National Institute for Petroleum and Energy Research in Bartlesville, Oklahoma, produced a report that includes several concepts that are relevant to this discussion (NIPER, 1986). It defines EOR as "petroleum recovery following recovery by conventional primary and secondary methods." No explicit mention of IOR is found in the public dictionary. Lake (1989) defines EOR as "oil recovery by the injection of materials not normally present in the reservoir."

Notice that these sources refer to EOR methods as a subset of processes—only one part of a greater list of improved oil recovery or IOR strategies. In this sense, the Society of Petroleum Engineers' IOR definition seems to imply that any strategy that aims at improving hydrocarbon flow toward wells or at recovering more oil after primary or secondary methods have reached their economic limits can be considered an IOR method. This definition is interesting because it does not necessarily imply that oil has to be stranded in the reservoir; instead, the definition resorts to continued production economically, after primary or secondary methods exhaust their ability to yield economic hydrocarbon production.

Well architecture, such as horizontal versus vertical wells, operating conditions, well spacing, and so forth, can affect oil production and recovery factors. The SPE's definition of EOR is somewhat more ambiguous from our point of view, but it clearly states that EOR engulfs processes that improve recovery after the primary production phase. We have included the SPE's definitions of primary or secondary recovery because they are relevant to this discussion. Notice that primary recovery relates to naturally supported production or natural drive by using the energy of the reservoir.

We should also point out that this production *does not include energy or fluid injection,* as an extension of the SPE's definition of primary recovery. It does not explicitly exclude assisted primary production with the use of pumps or artificial lifts, but net fluid or energy injection does not occur in this case.

Examples of assisted production include any form of gas lift, which in essence provides mechanical potential energy for lifting fluids but no energy to the reservoir. In systems such as these, gas is "bubbled" downhole to assist fluid lift; however, no fluid is actually injected into the formation. The secondary oil recovery phase includes pressure maintenance strategies as well as displacement methods such as waterflooding. The injection of energy and/or fluids is in agreement with the aforementioned definition. Lake's definition, as explained in his 1989 book, includes all recovery processes and many recovery agents. What we have in common with this definition is that EOR is not tied to a particular stage of production.

When we discussed development plans in Chapter 1 (see Figure 1.1), we indicated that EOR supplied a tertiary recovery method. However, we added that there is no technical or theoretical reason why an EOR method cannot be initiated at any stage of production. Our definition of EOR is "a set of production technologies that involve the injection of energy or fluids to improve oil recovery at any stage of production, whether primary, secondary, or tertiary, with the purpose of increasing the total recovery above what is possible through traditional methods—namely, primary or secondary methods (waterflooding and gas injection)."

You may wonder why unswept oil remains after primary and secondary methods are used. Remember that we are referring to the maximum economic limit of these processes. Although you already may have some knowledge of the reasons, we will still discuss the limitations of primary and secondary processes to provide background information on what most people want in the design of an EOR process. This also gives us the opportunity to introduce some concepts that are traditionally used in EOR.

2.3 ENHANCED OIL RECOVERY METHODS

This section is dedicated to the classification of EOR methods, so by grouping them, tasks such as screening can be facilitated. However, you should be aware that it is not always possible to group EOR processes and to use unified criteria for screening and decision making by referring to categories. Before we discuss the types of methods, let us examine recovery mechanisms and controls.

2.3.1 Oil Recovery Controls

First, this subsection is a brief detour in the style of the book to provide background material for understanding EOR methods. The question is, why are there limits to recovery in any oil recovery process and in particular the conventional recovery processes of waterflooding and gas injection? To answer this question, we need to familiarize ourselves with traditional wisdom on multiphase flow in porous media that is relevant to oil recovery processes.

The issue at hand is an injected agent that cannot completely dislodge the oil in the pore space. The lack of a complete sweep of oil in place involves a number of controlling mechanisms operating at both the pore and at the reservoir scales. The former mechanisms, the pore-scale retention mechanisms involve a number of interfacial effects that are generically associated with the rock–fluid interactions that lead to oil trapping or retention. The competing forces and mechanisms can be summarized by a few important dimensionless numbers. The latter mechanisms (i.e., at the macroscale) are associated with permeability heterogeneities, channeling or thief zones, or fracture networks, and a viscous ratio called the mobility ratio.

The model of oil trapping is referred to as "snap off" (Lake, 1989). To understand how water traps oil, which is the paradigmatic example, you have to recall that immiscible phases develop interfacial tension, σ_{ow}, at the interface between the two fluids. This means that when two or more immiscible phases come into contact, interfacial energy is created. This translates in turn into a tension or stress on the surface of the interface, just like a membrane or a balloon. As a result, work is required to deform the fluid–fluid interfaces. When the immiscible phases are located in the pores of a rock, the interfaces curve, and a pressure difference across the interfaces develops—namely, the capillary pressure.

You need just one more piece of information to help you complete the picture. The wettability, or relative ability of one fluid to wet a solid surface (pore surface) in the presence of a second one, determines the arrangement of the fluids in the pore space. If water, as presumed for a long time, wets a rock preferentially in the presence of hydrocarbon, then it will tend to sit on the solid surface while the hydrocarbon will sit in the inner portion of the pore space. Due to the presence of curvature in the pore space, water will choke the oil and disperse it into the rock.

In the case of water as a wetting phase, to dislodge the oil, you need to overcome the pressure difference between the oil and the water because for an oil blob to pass through a constraint, either the interface must curve or you must overcome the capillary pressure barrier. In theory, this trapping mechanism is responsible for the limited efficiency of water as a displacing agent, which leads to the concept of residual oil saturation or immobile oil fraction in the rock. The higher this saturation value (the more oil that stays trapped in the rock), the lower the recovery of oil.

To understand how the pore-level trapping of oil can be overcome, we need to write the first dimensionless number, the capillary number:

$$N_{ca} = \frac{v\mu}{\sigma \cos\theta}$$

The terms in the equation are velocity (v), fluid dynamic viscosity (μ), interfacial tension (σ), and the contact angle (θ). It suffices to say that θ relates to the wettability on specific surfaces. The capillary number reflects the ratio between two competing forces: the viscous drag of the fluid ($v\mu$) over the interfacial contribution given by interfacial tension (σ). If this number is small, fluid motion is impacted or dominated by capillary forces, while viscous forces dominate for $N_{ca} > 1$.

This conventional theory is the reason why you might want to add surfactants to the water, since these soaps can lower the interfacial tension (and change the wettability) to increase the capillary number and reduce trapping. In fact, it has been shown (Lake, 1989) that the residual oil saturation decreases when the capillary number increases beyond 10^{-3}. This is particularly true for sands and sandstone, but there is no critical capillary number for rocks like carbonate, although as a rule, the decrease in immobile oil saturation still decreases with an increasing capillary number.

Another important control on recovery is the mobility ratio. We would like to avoid equations. The intuitive idea conveyed by this ratio is that when a viscous fluid such as crude oil is displaced by a less viscous displacing phase, viscous fingering tends to occur, with a consequent reduction in macroscopic sweep efficiency. This is why polymers are added to the displacing water.

Oil recovery controlling mechanisms that were described earlier also apply to gas flooding. Although gas displacement efficiency is higher than that of water at the same reservoir pressure, oil saturation reduction cannot be drastically reduced at immiscible conditions. However, miscible gas (e.g., high-pressure gas) or solvent (e.g., propane or enriched gases) injection methods can improve oil recovery because they can lower residual oil saturations more than waterflooding alone.

Miscibility of the displacing solvent, or high-pressure gas, with reservoir oil results in the reduction of interfacial tension with a corresponding increase in the capillary number and oil recovery. However, because of the high mobility of gas and reservoir heterogeneities (e.g., presence of thief zones or high permeable channels), gas injection is preferred in light oil ($> 35°$API gravity) and low-permeability (< 50 md) reservoirs. In addition, gravity drainage represents the most effective gas-injection strategy (up-dip injection and down-dip production), which clearly exemplifies the impact of reservoir geology on oil recovery mechanisms by gas flooding.

Finally, reservoir pressure also plays a key role in oil recovery mechanisms. If reservoir pressure is below the bubble-point pressure (P_b), displacement efficiency is reduced due to the relative permeability effects caused by the multiphase flow of oil, gas, and water in the reservoir. The latter may also impact the feasibility of reaching miscibility with gas-injection methods, with the consequent lower oil recovery factors.

We will now examine some relevant EOR processes.

2.3.2 Classification of EOR Methods

The following is the widely accepted classification of EOR methods:

Thermal. This includes steam stimulation, or "huff and puff"; steam flooding; steam-assisted gravity drainage (SAGD); and in situ combustion or, in contemporary terms, air injection. Other current noncommercial technologies include electromagnetic heating from resistive heating at low frequencies to inductive and dielectric heating at higher frequencies, including microwave radiation.

Chemical. This family of methods generally deals with the injection of interfacial-active components such as surfactants and alkalis (or caustic solutions), polymers, and chemical blends. Surfactants for foam flooding come in several categories, including those intended for deep conformance in solvent flooding.

Miscible or Solvent Injection. These methods are frequently associated with a form of gas injection using gases such as hydrocarbon gas (enriched or lean), carbon dioxide, and nitrogen. However, the solvent, though not necessarily economic, can be a liquid phase. Supercritical phases such as high-pressure carbon dioxide are good solvents.

In modern enhanced oil recovery applications, coinjection of IOR or conformance agents, such as gels or foams, can be necessary. More recent developments include the injection of carbon dioxide-soluble surfactants to generate in situ foams for mobility control. Some EOR methods that have been extensively tried in the field include microbial-enhanced oil recovery that could fall in any of the aforementioned categories, but some of the mechanisms involved are not fully understood.

Thermal

As you might correctly surmise, thermal methods relate to processes that require the injection of thermal energy (Lake, 1989; Satter et al., 2008) or in situ generation. The paradigm of thermal processes is steam flooding. In heavy oils, the main mechanism sought is reduction of oil viscosity, which facilitates the movement of oil toward producers. The most successful strategy of steam flooding is the cyclic steam injection, or huff and puff (Satter, 2008; Thomas, 2008). In this method, steam is injected at high rates for a period of time, generally for weeks; then the formation is soaked for a few days by shutting the well and then putting it back into production. This is frequently used in heavy oils (10–20° API). In lighter oils, the application of heat leads to vaporized light oil fractions, which can act as solvent fronts.

This in fact is not unlike miscible displacement, but the mass transfer between the phases is somewhat different from gas-injection processes because steam distillation acts differently. Generally, after several cycles of cyclic steam injection, projects are converted into steam flooding as a strategy to maximize oil recoveries.

In situ combustion or air injection is often referred to as fire flooding (Mahinpey et al., 2007; Thomas, 2008). In this process, air or oxygen is injected to burn a portion of the oil in place. Depending on the oil, two basic modes occur: low-temperature oxidation (LTO) and high-temperature oxidation (HTO). The complex kinetics of cracking can lead to upgrading the oil in the reservoir (Mahinpey et al., 2007). More comments about this process are provided in the EOR status chapter. Other methods, such as steam-assisted gravity drainage (SAGD), are discussed in other sections. This type of method has been confined to the Canadian Oil Sands, despite their relative success.

Chemical

Traditionally, this method's target is the increase of the capillary number (Lake, 1989; Thomas, 2008). The best-known method is micellar-polymer (Lake, 1989). After significant technical successes in field trials, the process gave way to new alternatives, such as alkaline-surfactant-polymer (ASP) flooding, and a renewed interest in surfactant-polymer (SP) flooding. Straight polymer flooding has been a sustained production method in many areas, China being the most successful case (Satter et al., 2008).

In ASP, the polymer acts as a mobility control agent, while the alkali and surfactant act synergistically to widen the range of ultralow interfacial tension (10^{-3} mN/m). In SP, which is a combination of two surfactant (a surfactant and a cosurfactant) cosolvents, no caustic agent is used.

Miscible or Solvent Injection

This category of methods relies on the injectant's miscibility with the oil phase. The solvent is injected by flooding with one of the following:

Hydrocarbon miscible. The main mechanisms involve generating miscibility, increasing the oil volume, or swelling and decreasing the oil viscosity.

Carbon dioxide. The CO_2 flood leads to miscibility by extraction of oil fractions. It requires lower pressure than hydrocarbon miscible flooding. The mechanisms are similar to those of other miscible (e.g., vaporizing gas drive) processes.

Nitrogen and flue gas. Due to the high miscibility pressure, these processes have high capital expenditure, or CAPEX, and are seldom used. Vaporizing light oil fractions creates miscibility. The injection of these gases provides a gas drive mechanism.

As a result of the low viscosity of solvents, viscous fingering is a frequent problem with these processes. Also, override by the less dense phase leads to poor sweep efficiency. To mitigate these problems and reduce the solvent requirements, a process of alternating water and gas is used. This successful EOR strategy is called water-alternating-gas (WAG), and it is frequently used in carbon dioxide flooding to increase sweep efficiency and decrease the need for expensive solvents.

Simulations and Simulation Options

3.1 INTRODUCTION

Making predictions about reservoir performance is basically the same as making decisions about enhanced oil recovery (EOR) projects. Simulations have many purposes. Reservoir simulation is often used to forecast the production and reservoir conditions under specific development or exploitation conditions. The chapter discusses some of these potential applications. According to Mustafiz and Islam (2008), many approaches to predicting production performance have been developed over the years, including analogical, experimental, and mathematical.

We will focus only on the mathematical methods to illustrate their relevance to EOR, especially early in evaluation exercises. One traditional approach to this is material balance, which, as its name indicates, is based on mass balance. You can also think of this as a zero-dimensional approximation because a full description of the reservoir is not necessary. In addition, this approximation considers the storage capacity's time-independent representations (no changes in the porous media) to be relatively simple thermodynamic representations of the fluids that uniquely define the entire reservoir. This approach works well for many situations, including primary production stages and waterflooding.

Decline curve analysis is based on a constant form of the reservoir performance that assumes a decline in the performance curve (production) in any of the typical representations: exponential, hyperbolic, and harmonic. The constant operational condition is a necessary assumption for this method to be predictive.

Mustafiz and Islam (2008) refer to the statistical approach as the derivation of correlations from numerous reservoirs to serve as predictive equations for the reservoir of interest. We add that this approach requires identification of reservoir analogues, but it will not elaborate much along these lines, except to say that this is the basis for advanced screening. However, a significant difference from traditional approaches is that advanced screening requires multidimensional projections that allow one to establish analogies between reservoirs on the basis of more quantitative analyses. As Mustafiz and Islam point out, blind statistical analysis can be misleading if spurious correlations are built on baseless physical connections among variables.

Analytical methods are simplified representations of the principles used in numerical reservoir simulations (to refer to more complex simulation methods). These methods rely on symmetric geometries of well distributions or configurations of injector-producer pairs in confined reservoir geometries. Fluids are usually represented as incompressible; more complex thermodynamic representations might be possible but are seldom used. We talk more about analytical tools later in this chapter. We believe there is a place for this type of approximation in EOR evaluations, particularly when limited data is available or time is a severe constraint (see chapter on methodology).

3.2 A SIMULATION MODEL

A reservoir simulation model accounts for a number of conservation laws, including those described in the following subsections.

3.2.1 Mass Balance

This law's equations are typically written in volumetric formulations that allow the use of saturation as an independent variable. The so-called black-oil simulators essentially account for three phases: oil, water, and gas. There is a possible mass exchange between the oil and the gas that is controlled by the pressure–volume–temperature (PVT) properties. In more complex simulation models, components or, more precisely, pseudo-components are conserved. In multiple-contact miscibility, the exchange of oil fractions leads to reduced interfacial tension and enhanced sweep efficiency. When you consider a waterflooding problem in heavy to light oils, a black-oil formulation may suffice.

Recovery processes that rely on mass-transfer mechanisms (e.g., enriched-gas, N_2, or CO_2 injections) require the use of some form of compositional balance in a compositional simulator. Intermediate approximations, such as a solvent model in black-oil simulation, are often used as a result of elevated computational costs of compositional simulations. Reactive flows, as in the case of air injection projects, also require some form of compositional interpretation. Other more restricted forms of compositional simulations that are based on empirical partition coefficients are adopted in thermal simulations to include pseudo-components, such as "heavy oil," "light oil," and so forth, in addition to thermal effects.

3.2.2 Momentum Balance

This law is represented by Darcy's law, generally in its multiphase flow representation, including relative permeability curves and often capillary pressure curves. Corrections to nonlinear effects are included through concepts such as the skin factor.

3.2.3 Energy Conservation

This law is especially relevant to the simulation of thermal processes. Energy transfer in steam injection—even in the form of steam stimulation or huff and puff—is paramount to the performance of this process. In fact, if excessive energy losses occur through overlaying strata, then this might be a deterrent to a steam injection project.

3.2.4 Pressure–Volume–Temperature

The PVT properties of fluids and rock are intended to account for the thermodynamic conditions and the phase changes in the simulation. In black-oil simulations, PVT models are relatively simple and oil can be treated as a phase in equilibrium with the gas—for instance, in flash tests. Compositional PVT is necessary for more complex fluids and in particular to account for miscibility conditions between injectants and native reservoir fluids. The multiple-contact miscibility is the result of a mass transfer among phases to reach miscibility. Vaporizing-gas drive, for instance, might require going beyond traditional thermodynamic representations.

PVT also applies to the rock through compressibility of the matrix. In recent developments, geomechanic coupling between fluids and rock have been attempted to account for a myriad of effects that cannot be modeled with approximations that decouple the fluid behavior and the rock, such as subsidence. Water, on the other hand, can often be represented with simple compressibility functions.

One aspect of the coupling between fluid pairs is how pressure—say, between water and oil—links two phases. Capillary pressure usually does this, but other conditions might appear that require an explanation. Of course, you cannot carry out a simulation if the initial and boundary conditions and the operational conditions are not provided.

3.3 OBJECTIVES OF SIMULATIONS

Simulations in reservoir engineering have several objectives, which include the ones described next.

3.3.1 Reservoir Characterization

After geoscientists and engineers have established the main characteristics of a hydrocarbon trap, a number of uncertainties must be handled resulting from limitations either in the data sources or in the ability to establish the geometry and properties of the reservoir. In this sense, simulation models can be used to determine how compatible the information about the reservoir image is with the performance measurements of the reservoir. We often start with a geologic model that is consistent or compatible with our interpretation of the constraints—for instance, the tops of the formation (geophysics and petrophysics), understanding the sedimentation process (sedimentology and geology), and the storage capacity (petrophysics), among other aspects. This is the usual static model.

A simulation model might make it possible to determine if production or tracer data or other forms of dynamic data are compatible with the reservoir image. Streamline simulation is an effective tool for evaluating waterflooding projects, including well balancing, reservoir connectivity, and so on (Alvarado et al., 2002).

3.3.2 Production Forecasts

This type of simulation is usually based on "fitting" (history matching) the production data (e.g., cumulative oil and rate, watercut, water production), the bottomhole pressures, and softer data such as timelapse seismic (Artola and Alvarado, 2006). We refer you to specialized references on reservoir simulation for details. Be aware that different software companies provide equivalent simulation tools; however, some are more effective for specific simulation problems. For instance, we find

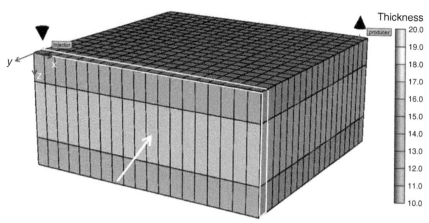

FIGURE 3.1   A quarter of a 5-spot model for water and ASP flooding simulations.

that some of them are industry standards for chemical and thermal simulation, while others are better tools for compositional simulation. Another important consideration is how easy the tool is to use.

Figure 3.1 shows an example of a simplified grid model. The figure represents a quarter of a 5-spot model, which is frequently used as a sector model in exploratory simulations. In a particular analysis, we wanted to compare the forecast of an alkaline-surfactant-polymer (ASP) design for a heavy-oil reservoir (unpublished) with waterflooding. Data from core floods were used to establish the simulation inputs, as shown in Figure 3.2 (relative permeability curves).

For this case, a commercial simulator based on partitioning was used. This particular piece of software assumes smooth changes in relative permeability curves as the capillary number increases as the result of surface-active reagents such as surfactants. Waterflooding curves are shown in Figure 3.3.

3.4 ANALYTICAL SIMULATIONS

Because analytical simulations are so relevant to the methodologies discussed in this book, we will examine some of the issues associated with them here. Figure 3.4 illustrates the main idea for using the analytical simulators described. Most of the discussion can be found in publications by Alvarado (2001) and Alvarado and colleagues (2003). This section illustrates three important points:

1. It is possible to sketch maps that represent variability on a reservoir, even with simple tools. This conforms to the Moving Mosaic technique (Balch et al., 2000; Hudson, Jochen, and Jochen, 2000;

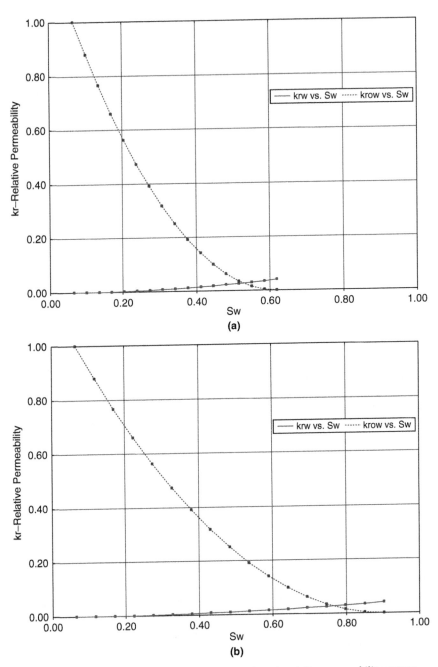

FIGURE 3.2 Low (a) and high (b) capillary numbers in relative permeability curves.

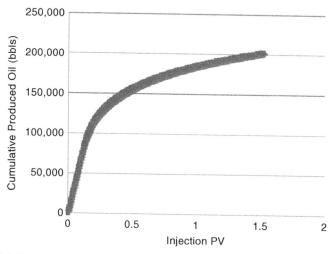

FIGURE 3.3   Cumulative oil production for waterflooding from the model in Figure 3.1.

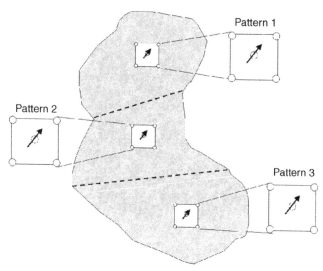

FIGURE 3.4   A hypothetical reservoir where the different well patterns represent the property variability over the area.

Hudson, Jochen, and Spivey, 2001; Voneiff and Cipolla, 1996), which is a method to find "production potential maps" (Alvarado, 2002).

2. Analytical simulators that can handle several EOR processes, even in a limited fashion, must be available. You cannot resolve all of the required physics of multiphase flows for some complex processes using analytical solutions, but you may be able to draw a rough image of the performance of an EOR process. This, in fact, may be all you need to build a decision framework and proceed to support a decision.

3. The possibility of carrying out a history match is important. This can be shown by IDPM, which is a streamtube model to simulate infill drilling in 5-spot patterns (Fuller, Sarem, and Gould, 1992).

Later in this chapter, we discuss some of the limitations of analytical simulators. In PRIze (ARC, 2006; PRI, 1995), one of several simulators available, a history match for waterflooding can be achieved by tuning the Dykstra-Parsons coefficient (Craig, 1993).

Waterflooding is often the basis for several EOR processes, especially in the case of chemical methods. Similar approaches are followed for the simulation of other EOR strategies. Several techniques are used for waterflooding (Thakur and Satter, 1998), as well as for the processes that follow:

Volumetric method. Very much derived from material balance (reservoir analysis). It is convenient for estimating ultimate recovery.

Empirical methods, based on correlations. These are the same as some of the statistical methods mentioned before.

Classical methods that include the Dykstra-Parsons method or more elaborate ones such as the Buckley-Leverett method. Table 3.1 shows the major limitations of several of these methods.

Performance curve analyses. The most common analyses are logs of production rate versus time, production rate versus cumulative production, and water cut or oil cut versus cumulative production.

Simulations. These include streamtube and streamline simulators and finite-difference, finite-volume, or finite-element black-oil or compositional simulators. When using streamtube models in simplified geometry and representations of a reservoir section (pattern), streamtube models can be considered analytical simulators.

Knowing the limitations associated with each approximation (refer to Table 3.1) is indispensable when you want to determine the representation of the recovery process in analytical simulators. This avoids misinterpretations of the simulation results and incorrect predictions.

3.4.1 Steps for Analytical Evaluations

One possible workflow for analytical simulations is particularly helpful, although you may have to adapt it for your own needs. It uses two bounding approximations for analytical simulation: the Dykstra-Parsons model or zero vertical segregation (Figure 3.5) and vertical equilibrium (Figure 3.6).

You should always use formulations that you understand. By establishing expected reservoir performance behavior boundaries, you will gain insight into possible critical conditions. The steps to use follow (see page 26).

TABLE 3.1 Characteristics of Commonly Used Analytical Approaches

	Dykstra-Parsons	Stiles	Prats et al.	Buckley-Leverett	Craig et al.
Flow: Pistonlike	X	X	X		
Flow: Frontal				X	X
Crossflow: Areal	No	No	No	No	No
Vertical	No	No	No	No	No
Initial gas saturation	No	No	Yes	No	Yes
Sweep: Areal	No	No	Yes	No	Yes
Sweep: Vertical	Yes	Yes	Yes	No	Yes
Mobility ratio	Yes	1.0	Yes	Yes	Yes
Stratification	Yes	Yes	Yes	No	Yes
Pattern: 5-spot	No	No	Yes	No	Yes
Pattern: Other	No	No	No	No	No

Source: Thakur and Satter, 1998.

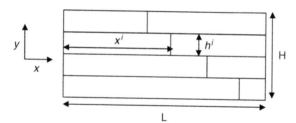

FIGURE 3.5 The Dykstra and Parsons layering scheme. Layers are reordered from shorter to larger sweep times.

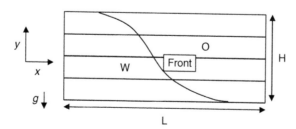

FIGURE 3.6 A vertical equilibrium approximation. Water segregates underneath the oil.

Step 1: Define Clear Objectives

Although this step is often omitted, it is critical that firm goals be set before proceeding with the next step. If the evaluation is going to be performed by a team, influence diagrams and other tools should be used to minimize misunderstandings. Analytical simulation can sometimes make people feel as if they have to rush the evaluations. This implies that the framing phase should precede the simulation exercises.

Step 2: Collate the Data

It is a good idea to use the least amount of data possible. Information on sedimentology, for example, would be the most useful to estimate or to determine the pay continuity and the stratification scheme in the reservoir model. This does not mean that you need an entire model from integrated studies, but just a trap structure that is enough to generate gross rock volume (GRV) or the top and bottom of the reservoir; this must be available to establish local simulation models for the analytical simulator. Petrophysical data—local well information—can be a starting point to generate local models, as is often done in reservoir characterizations. Net reservoir data are important to assign the correct reserves.

Step 3: Screen the EOR Methods or Reservoir Pairing

This step is carried out by binary criteria (Yes/No) or a look-up table, data mining, or some other method. This can be done in two steps. During the first step, prior to more extensive data gathering, the average properties of the reservoir are used to reduce the number of EOR methods. However, binary criteria cannot be used to rank methods for a given reservoir.

Step 4: Use Project Critical Parameters

As we have already seen, operational issues and local restrictions can involve two categories of critical project parameters, but distinguishing the two is not important at this stage, except for preliminary screening. For instance, a lack of injection gas may disqualify gas flooding, but it does not technically eliminate the process. This is what is meant by keeping the objective clear. What is important at this stage is determining the critical reservoir uncertainties and the process technical variables. At this point, variables such as pay continuity should be scrutinized. Once a reduced list of processes or a given process is selected, a more detailed screening should proceed (step 5).

Step 5: Section the Reservoir

This step has two parts and is perhaps the most time-consuming one. First, geological features such as important faults should be identified so that isolated blocks can be treated separately. Then "applicability maps" should be sketched. These maps are the result of analyzing CPPs on the reservoir map. For instance, clay content can be a critical parameter for most chemical methods because the mineral content of the reservoir rock dominates adsorption.

Figure 3.7 illustrates the spatial differences in isopach maps in the case of the La Salina project. The clay content distribution can be used to section the reservoir, along with the water saturation maps, since these two variables have considerable impact on the chemical methods. This approach is valid for other processes as well, but then other variables become critical. For instance, in the case of gas flooding, if the miscibility condition varies spatially in the reservoir, the recovery efficiency will not be equivalent in all sections of the reservoir, and overall performance may not be well represented by one well pattern.

In the VLE area, for example, a section of a light-oil reservoir selected as a target for EOR—specifically water-alternating-gas (WAG)—is a good example of the significant pressure variations (see Alvarado, 2001, and references therein). The WAG process in that field is immiscible. Figure 3.8 shows the pressure distribution in separate blocks of the VLE field. (We will come back to this case to illustrate analytical simulations in more detail.)

What is most relevant is to derive heuristics to determine a few areal sections in reservoirs but not to refine too much because regional statistics (a few well patterns) are necessary for the whole process to operate.

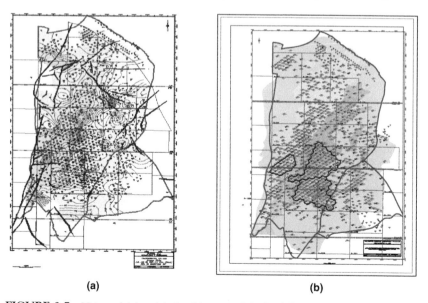

(a)   (b)

FIGURE 3.7   Net-sand (a) and facies (b) maps of the La Salina reservoir.

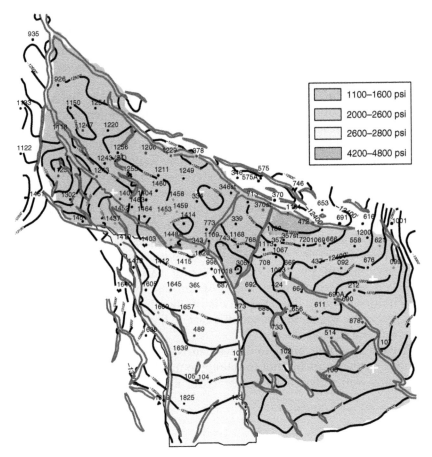

FIGURE 3.8 Pressure map for the VLE area.

Finally, allocation of local average well patterns is required. In this case, it is assumed that a pattern, along with its local properties, is representative of an area of the reservoir. Properties such as oil or water saturation have to be similar enough in an area, or a proper (volumetric average) must be taken.

Step 6: Match the Local History

In this step, the previous production history or well performance is used to estimate pay continuity and kh values. Biglarbigi and colleagues (1994) identified two methods for modeling pay continuity and thus carrying out a history match. The "net pay" method is preferable because it is more physically sound and can be handled easily in most simulators. Stream-tube models (e.g., IDPM) can manage vertical heterogeneity directly or with a Dykstra-Parsons (DP) coefficient. The SWORD Prediction module, part of the simulator SWORD (IRIS, 2007), can be used in a similar way for this purpose. PRIze, except for waterflooding, uses the D-P coefficient only because stratification is not explicitly involved.

Step 7: Perform a Sensitivity Analysis

A deterministic analysis should be performed, avoiding violation of independence of variables. The example provided before was pay continuity and net pay. Those two are intimately related. Tornado diagrams can be generated at this stage (Skinner, 1999). Also, this stage should serve to revise critical parameters for the analytical simulation.

Step 8: Perform a Decision Analysis

At this point, all of the information obtained and the outcomes from the simulations should give you a sufficient foundation to proceed with a decision. The type of process that follows is intimately linked to the objectives of the project. If uncertainties are getting in the way of proceeding to the next step, perhaps a value-of-information (VOI) analysis should come next to decide if more characterization is necessary.

Some of the preceding steps may be skipped or simplified, but a few more can be added as well to accommodate other objectives. This workflow has been presented as it was originally conceived so you can use it independently if you prefer. However, we recommend making progress with the strategies delineated in this book and using them as you find them appropriate.

3.4.2 Field Case

The VLE project (Alvarado, 2001) is used here to illustrate the previously described steps. It was designed to evaluate the immiscible WAG process in deep reservoirs in western Venezuela. One of the sources of data was a full field model of the VLE. However, only limited information was obtained from it, in particular on a few optimized waterflooding patterns proposed as an alternative to WAG injection. Two areas of the reservoir are analyzed here.

Data from the full-field case are processed to generate analytical models. Certain details of the completion scheme cannot be analyzed fully, but as much realism as possible is incorporated. The SWORD 1.0 prediction module was used for this example. This module was designed so injector-to-producer sections can be simulated. We use the same steps as in the analytical simulation.

Step 1: Define Clear Objectives

As a preliminary exercise, an informal survey was carried out with part of the team working on the VLE project. Questions on the status of the Integrated Laboratory Field (IFL) were posed to gain insight and to give more meaning to the exercise for this field. As far as the results of the WAG pilot test were concerned, it helped to improve oil production and to control water and gas production. However, the tracer program's results yielded a few surprises in terms of reservoir continuity between injector-to-producer pairs.

Gas supply for injection is rather scarce, which is an important soft issue for this case. Waterflooding optimization was considered a mid-term alternative to WAG injection. From there, it was decided to study the performance of the waterflood in a couple of areas of the reservoir. The objective was to determine if the undercurrent conditions had significant differences in potential from the waterflooding in the field. The results were expected to show the differences in potential between the two regions of the reservoir.

Step 2: Collate the Data

As discussed, we have to collect a minimum of information to build analytical simulation models. Stratigraphy, reservoir maps, and production histories were obtained from several sources. They mainly came from presentations, but published information was used as well. The four overlying reservoirs in the C2 Eocene Unit of interest, because they relate to the IFL project, were C-20, C-21, C-22, and C-23. The C-23 unit was the main focus of interest for the WAG project. Layering information was taken directly from sections of the model used for a master's thesis project (Stirpe, 2003). This, of course, is not what is normally done in an evaluation of this sort, but we will keep data to a minimum to simulate a scenario where necessary information is scarce.

The relative permeability and PVT data were obtained from the model, and they were assumed to be correct, despite the academic nature of the data source. It has been reported that relative permeability is reliable at high water saturation, except at the endpoint. Saturation, porosity, net-to-gross values, and layer thickness are all taken directly from the full-field model.

Step 3: Screen the EOR

Waterflooding has been ongoing in this field for some time, with relatively good success. Several reservoirs are still subject to water injection, although watercut in some wells is quite high. The C2 Eocene Unit reservoir that is under investigation here has been evaluated as a candidate for the WAG process. Gas injection is discarded, not because of technical reasons as far as the process goes, as we will see later, but for the lack of enough injection gas. Perhaps this will turn out to be a simple screening exercise because the field history has already proved the viability of waterflooding and gas flooding to the processes in this field, but the purpose is to illustrate a recommended line of thought.

Cyclic water injection has been simulated in the pilot area, and the results show potential for waterflooding optimization (Alvarez et al., 2001). The reservoir represents a highly stratified environment, with limited vertical communication and enough remaining reserves to make the process possible. The facilities for water injection possess enough remaining capacity to deal with the new IOR process, but it is still under scrutiny.

Step 4: Establish Project Critical Parameters

Given the extended experience with waterflooding worldwide, it is relatively simple in this case to identify several of the critical parameters. For illustration purposes, we have selected uncertainties in relative permeability. In addition, water and gas saturation, to some extent, can affect the waterflood, since the field has been under water and gas injection for quite some time. The area for waterflooding optimization has been shown to have an irregular water sweep. The latter means that layer continuity can play an important role in the waterflood. Finally, vertical communication can make a significant difference in waterflooding performance, although it will not be studied here, and it is assumed to be noncritical. The quality of the injection water is not questioned at this stage, since this issue was resolved early on in the design of the initial waterflood.

Step 5: Section the Reservoir

The chosen reservoir is a good example, with clearly identified geological features that allow us to divide it into distinct blocks. As was shown in Figure 3.8, the reservoir exhibits four areas where the pressure values allow us to conclude that the known major faults are effective seals—in other words, they are flow barriers. The high-pressure region toward the east is small enough to have no significant impact any analysis that is involved here. Production history indicates that even more compartmentalization may exist in the reservoir. At this stage, however, it is not necessary to do any further reservoir sectioning. Because the progress of the aquifer is structurally low in the field, no additional potential is expected from those areas.

The northwest block had enough remaining reserves to make it a good candidate for analysis. Figure 3.9 depicts this block, which is taken from the full-field simulation model and can also be seen in Figure 3.8. Two rectangles highlight the areas that were selected for evaluation. They also correspond to the areas selected for water optimization in the full-field simulation. They are considered as representative of two different areas of the block: structurally high and low (to account for high gas saturation and high water saturation, respectively).

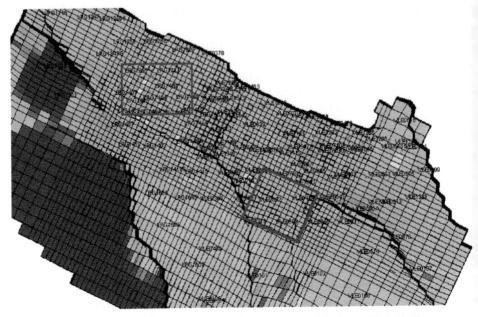

FIGURE 3.9 Section of the northwest area of the full-field model of the VLE field. The two red rectangles indicate the selected areas for analytical simulation.

Figures 3.9 through 3.13 show the zoomed-in areas of the selected block, along with 2-D sections between the injector–producer pair in each area. Intervals that offer little potential for production under water-flooding because of high water saturation will be shut in for the simulations. We can see that the cross-section between wells VLE-0692 and VLE-0708 (Figure 3.11) exhibits higher water saturation in lower intervals than that seen in Figure 3.13. Sensitivity analysis was carried out in these 2-D sections to illustrate the types of results that can be obtained with the SWORD 1.0 IOR prediction module.

Step 6: Match the Local History

In the VLE case, no explicit history matching was carried out, but instead a production history was used to corroborate information on lateral reservoir continuity and typical water and gas saturation in the block. Water injection was initiated with a pilot test in 1963, but the peripheral water injection project was not officially started until 1967. Gas injection, on the other hand, has not followed the initial plan, and water injection has been increased with time to make up for the lack of pressure support from gas injection. Irregular progress of the water displacement front, as well as channeling, has proved the heterogeneous nature of the reservoir. The latter motivated the request for infill drilling with a well spacing of 300 meters.

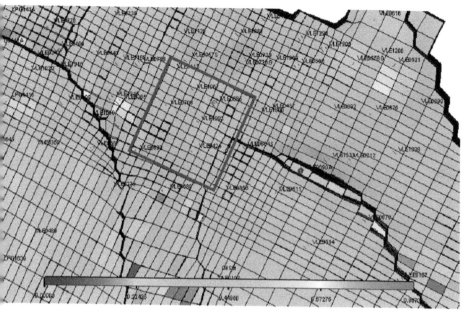

FIGURE 3.10 Closeup of bottom area of Figure 3.9.

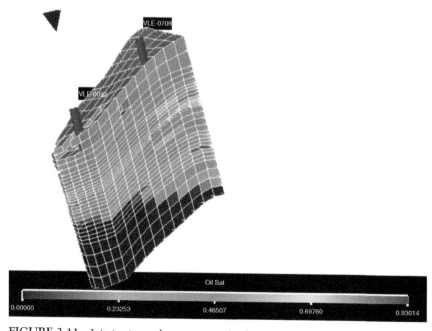

FIGURE 3.11 Injector-to-producer cross-section between wells VLE-0692 and VLE-0708 (see Figure 3.10). This cross-section is representative of the northernmost area under analysis.

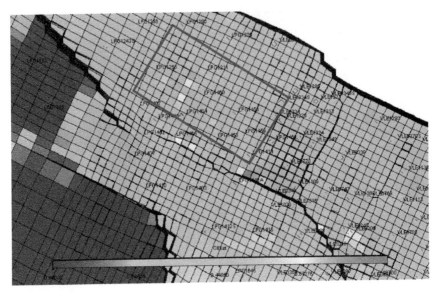

FIGURE 3.12   Closeup of top area of Figure 3.9.

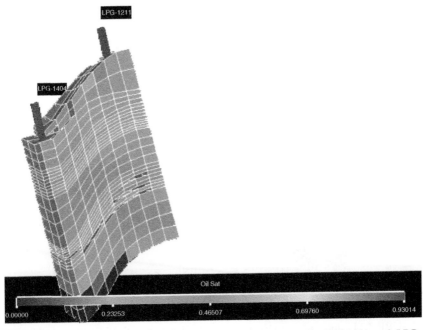

FIGURE 3.13   Injector-to-producer cross-section between wells LPG-1404 and LPG-1211. This cross-section is representative of the southernmost area under analysis.

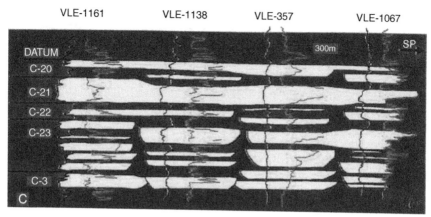

FIGURE 3.14   A cross-section of the reservoir that illustrates its current heterogeneity.

Drilling of infill wells at 300-meter spacing has shown very little interference with neighboring, older wells. The current view of the reservoir continuity, as shown in Figure 3.14, reflects the latter quite well. Sedimentology shows that the environment of the reservoir is comprised of bars and fluvial channels.

We can see in the figure that C-23, the lowermost reservoir, is expected to be continuous just below a 300-meter interwell distance, since the bodies have a width in that size range. The situation for the uppermost reservoirs seems to be somewhat different, perhaps with better lateral continuity, but in all cases it depends on the orientation in the field. To make the results comparable, both of the sections are oriented in the same direction. To model lateral continuity, a drop in reservoir continuity can be modeled as a known function of distance. In the case of VLE, lateral continuity could be set close to 90 percent at a 300-meter well separation. This number is only referential, and simulations should be sensitized with respect to it.

Step 7: Perform a Sensitivity Analysis

Most of the results at this point correspond to cross-sections that go from injector to producer, as was shown in Figures 3.11 and 3.13. The layering scheme in the simulator needs fixed horizontal and vertical permeability, porosity values, net thickness, and initial water and gas saturation. Also, all of the data that refer to the recovery factor are computed not from the OOIP but instead from the current oil saturation. As described in the section on analytical simulators, the total liquid production equals the total injection rate. In this case, fluids are taken as incompressible. A fraction of the planned injection rate in the 5-spot pattern is used for the cross-sectional simulation. In this case, 500 bbl/day were injected. Only DP calculations are shown. Simulations carried out with vertical equilibrium approximation yielded a warning on the validity of the approximation. From the survey, it was found that a fixed Kv/Kh value equal to 0.1 is considered for the whole field.

Although the completion scheme of the wells varies from well to well in the selected areas, the full interval is simulated. Unless specified, a continuity equal to 1 is used in the simulations. The first simulation corresponds to a sensitivity analysis on the cross-sectional width for a fixed water injection rate. From the information gathered, it is known that as much as 10,000 bbl/day of water have been proposed in some patterns in this field. A fraction of this injection volume, 500 bbl/day, has been used for the simulation. Figure 3.15 depicts the behavior of the oil recovery factor (calculated from the oil currently in place) as a function of time in a period of 20 years. Three cross-sectional widths were tested: 60 feet, 120 feet, and 180 feet.

Figure 3.16 shows the oil production rate for the same parameters employed to produce Figure 3.15. Three cross-sectional widths were tested: 60 feet, 120 feet, and 180 feet, as shown in Figures 3.17 and 3.18. The widest section (180 feet) yields a production profile that is reasonable, considering that production wells are shut in once the oil rate goes below 50 bbl/day in the field or the water cut is greater than 95 percent.

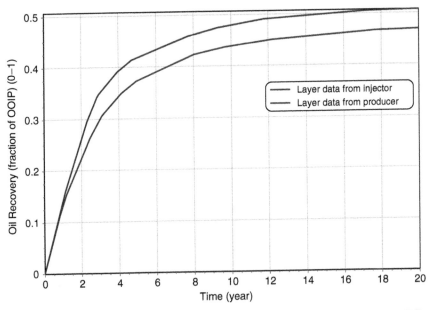

FIGURE 3.15   The recovery factor for a 2-D section between wells VLE-0692 and VLE-0708. The net pay in the 2-D section was assigned from either the injection well information or from the producer.

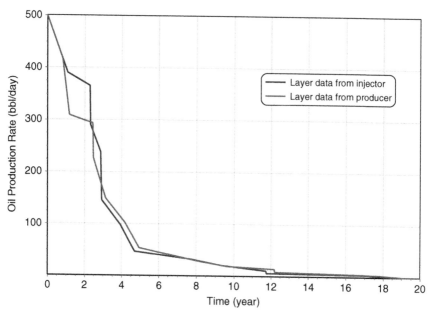

FIGURE 3.16   Oil production for a 2-D section between wells VLE-0692 and VLE-0708. The net pay in the 2-D section was assigned either from the injection well information or from the production well.

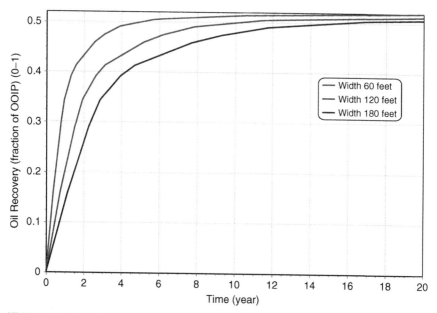

FIGURE 3.17   The sensitivity analysis for a 2-D section between wells VLE-0692 and VLE-0708. The water injection rate is 500 bbl/day, and three different widths were tested: 60, 120, and 180 feet.

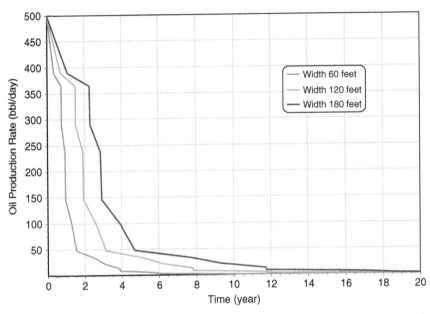

FIGURE 3.18   The sensitivity analysis for a 2-D section between wells VLE-0692 and VLE-0708. The water injection rate is 500 bbl/day, and three different widths were tested: 60, 120, and 180 feet.

Both conditions are approached after the sixth year, but this is sufficient for the purposes of these simulations.

All of the results presented from here on will use 180 feet for the cross-sectional width (similar results could have been accomplished with a lower injection rate and the same channel width). It is important to note at this point that, given the water and gas saturation observed in the producer, endpoint relative permeabilities had to be adjusted in one or two layers to satisfy the correct mass balance. Figures 3.15 and 3.16 showed that the recovery factor and the oil production rate were not too sensitive to whether properties are assigned from either the injector or the producer sides, although some differences are obvious.

This is probably so because the cross-section selected in this direction over the reservoir has good properties. Some of the neighboring wells exhibit little to negligible net sand in the same intervals in this cross-section. The latter reflects the highly heterogeneous character of the reservoir. In any case, the behavior of the oil production rate, and thus the recovery factor, should turn out to be between those obtained from the extreme cases. On the other hand, no effect of discontinuity has been taken into account up to this point.

Relative permeability has been identified as a critical parameter for waterflooding; this also true of many enhanced oil recovery processes). Endpoint relative permeability from the experiments exhibits poor confidence. Figure 3.19 shows cumulative production of the same 2-D section for three values of Sorw, while Figure 3.20 shows the oil production rate for the same conditions. Large differences in recovery should be expected, as demonstrated in Figure 3.19. Even in this simple analytical model, the effect of relative permeability is clear. This implies that a high-risk factor is the assumption that the same relative permeability curve holds for the whole field.

Simulations are performed using Sorw = 0.25 from here on, unless something else is explicitly specified. In the C2 reservoir, there are several intervals of interest, which are isolated from the rest by impermeable, zero-porosity layers (shale). By looking at the oil production profile, water cut, and injection rate per layer, we can quantify the contribution (importance) of each layer (not shown).

One issue that has not been dealt with thus far is the fact that average saturation can change greatly from the position of the injection well to the producers. In the case of VLE-0692, the aquifer has reached the lower interval of the reservoir, since it is located down dip. This situation may have a downside in the simulations performed when those intervals are

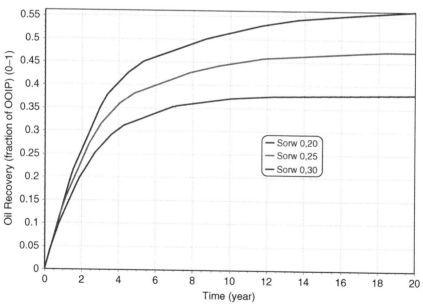

FIGURE 3.19 The recovery factor for a 2-D section between wells VLE-0692 and VLE-0708 and the effect of endpoint relative permeability.

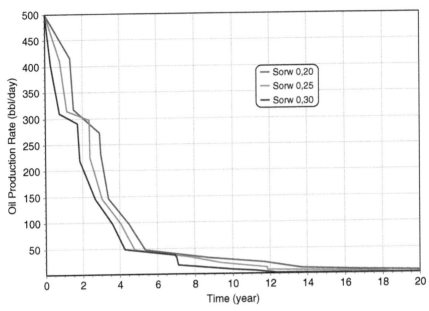

FIGURE 3.20   The oil production rate corresponding to Figure 3.19.

opened. The reason is that, although it may be convenient from an injectivity point of view, it will lower the amount of movable remaining oil in place. The latter has to be kept in mind when deciding on production from this area. The area up dip, where well LPG-1404 is located, presents lower water saturation, which is a less risky situation. Now a comparison between the two cross-sections is outlined. Given the results so far, not all of the sensitivities will be performed. Runs in both cross-sections are performed with the same parameters (180-foot width, 500 bbl/day of water injection, Srow = 0.25, and the same endpoint values).

Figures 3.21 through 3.23 show the production rate, the recovery factor, and the water cut. From the three previous figures, it should be evident that the potential of the LPG-1404–VLE-1211 section is inferior to that of VLE-0692–VLE-0708. There are several possible reasons for this. Recalling the fluid injection history in this reservoir, the crestal gas injection was responsible for displacing the oil up dip—that is, structural heights. Under a scenario of water injection in that location, very poor performance should be expected from layers that are saturated with gas; in the LPG-1404–VLE-1211 section, that involves most of the top layers.

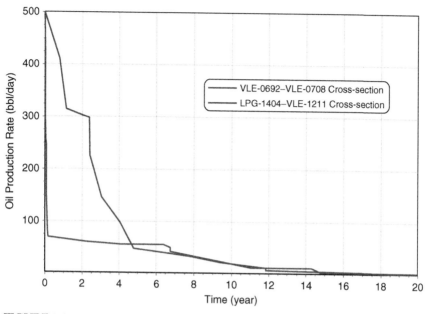

FIGURE 3.21   The oil production rate and a comparison between the two cross-sections.

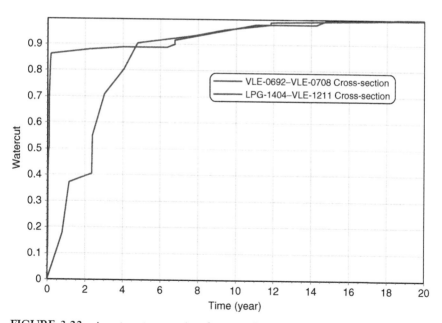

FIGURE 3.22   A watercut comparison between the two cross-sections. All simulation parameters are the same.

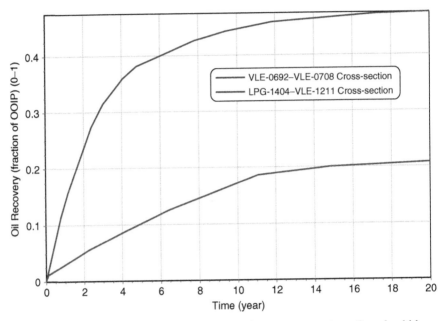

FIGURE 3.23 Cumulative oil production from the two cross-sections. Care should be taken because the values of the current oil amount in place are not the same.

Step 8: Perform a Decision Analysis

We will elaborate on influence diagrams later for this purpose. In our case, it should be clear that for success in terms of oil recovery, a completion scheme will be the most important decision, followed by the water cut limit, before shutting in a well. Other considerations could become important in a more thorough exercise.

In the larger context of the methodology, some redundancies can be observed. This is a necessary consequence of the implied flexibility of this type of approach. We insist that not only is flexibility important, but it is also indispensable when it comes to managing uncertainties.

4

Screening Methods

4.1 INTRODUCTION

This chapter examines the elements of screening that are required to make progress with decision making in the framework of our proposed workflow. Three screening styles must usually be combined to paint a good picture of the enhanced oil recovery (EOR) decision problem and to make rational progress. The first one, conventional screening, is the one most engineers are familiar with, and it is usually carried out by comparing average reservoir properties with data in a look-up table that contains validity limits for each parameter considered important.

Geologic screening is a way of looking at the reservoir type in terms of heterogeneity, connectivity, and other geologic characteristics that have been found to be important in managing risk or that correlate with process performance. Advanced screening helps when looking at possible combinations of variables and are sometimes referred to as multidimensional maps (to see more than three-dimensional projections). These projections are useful for finding proper reservoir analog.

This chapter is divided into three main sections, each of which explains a type of screening, along with examples.

Enhanced Oil Recovery
DOI: 10.1016/B978-1-85617-855-6.00010-3

4.2 CONVENTIONAL SCREENING

The most commonly used approach to selecting recovery processes for a reservoir is so-called conventional screening, which we refer to as "go–no go" screening. This strategy is based on look-up tables where intervals of validity are established on the basis of engineering considerations by collecting "expert opinions" or by analyzing data from successful field cases. A combination of all of these approaches is the most likely situation encountered. In this screening method, typically average representative fluid and reservoir properties of a particular field under evaluation are compared with intervals of the look-up table to decide whether the field or reservoir is suitable (which is why it is called go–no go) for a given recovery process.

Screening methods of this sort are well documented in the literature (Taber et al., 1997) or are available in commercial analytical tools; for instance, PRIze implements a direct look-up table strategy, while Sword (IRIS, 2007) relaxes the look-up table, using fuzzy logic to generate an indicator between 0 and 1 and thus allowing hierarchical selection of the process type (waterflooding, gas injection, thermal methods, and chemical processes).

An important consideration of look-up tables is that biases frequently arise because engineering considerations or experts' opinions are introduced in the process. For instance, PRIze was developed by the Petroleum Research Institute (formerly known as PRI; it is now part of the Alberta Research Center, or ARC; ARC integrated with Alberta Innovates, a new organization in Alberta), and as a result EOR applied to heavy oil substantially influenced expert opinions and sources of data. Sword, which was developed in Norway, is biased toward the Norwegian sector of the North Sea, where light oils dominate. The main goal of the screening analysis is to identify whether a specific EOR technology has been implemented under fluid and reservoir properties similar to those of to the field under evaluation.

As we begin our discussion of the methodological approach or the workflow at the core of our decision-making process for EOR, an alternative "conventional" screening that we developed will be explained through examples. This screening approximation is, as in other conventional screening approaches, based on a comprehensive comparison of reservoirs and fields under evaluation with an extensive database of international IOR/EOR of approximately 2,000 projects, as well as publicly (e.g., U.S. DOE Toris and Heavy Oil databases) and commercially available (e.g., SPE and the *Oil & Gas Journal's* EOR surveys) data sources. A bit of history might be helpful at this point.

The authors have collaborated throughout a decade or more on EOR screening, analyzing field cases, recommending enhanced oil recovery solutions, designing or evaluating laboratory experiments for EOR, predicting field performance through a variety of simulation exercises, designing pilot tests, and becoming part of integrated teams for asset development plans. As a result of some unique conditions, with limited field data, in searches for analog fields the world over have allowed us to collect a large database of field experiences that has become a source of "wisdom."

A number of talented colleagues have contributed to increasing the value of this database, although at present the task of maintaining or enhancing the database falls on the two authors of this book. We see this as a collection of experiences that allows anyone with the right focus and expertise (and we hope you are or will become one of those people) to mine this experience to improve decision making in EOR problems. In a nutshell, to explain the typical situation, most people focus on data, but you should at least concentrate on information. The sequence goes as follows:

$$\text{Data} \rightarrow \text{Information} \rightarrow \text{Knowledge} \rightarrow \text{Wisdom}$$

Most of us are not good at the data level, and computer systems can do this better than analysts. However, as we go up the complexity and synthesis ladder, we become better. The objective of the screening exercise is to rapidly see if a field or reservoir under consideration presents enough commonalities with field experiences in the same area or elsewhere. If the answer is positive, then the likelihood of finding referential information as to what the course of action was in similar reservoirs can be investigated; if, on the other hand, the reservoir under evaluation turns out to be an exceptional case with no comparable field conditions in EOR, care must be exercised to avoid excessive risk in the application of an EOR process.

Figure 4.1 shows the basic workflow of the conventional screening followed in the methodology proposed in this book. Operational, XY (Figure 4.2), and radar plots of average reservoir and fluid variables are generated to allow engineers (analysts) to preliminarily identify fields with similar properties. In this phase, engineers can determine a lack of published field experience for those methods in fields that have reservoir properties similar to those of the fields that are being evaluated.

The choice of variables is guided both by the availability of data and by the "intuition" (experience) of the analyst. Radar plots of six or more variables are also used to identify trends and ranges of preferences for (or the applicability of) a particular EOR method in multiple reservoirs before using more advanced screening methods (Manrique and Pereira, 2007). In this book's authors' experience, more than six variables might be required to identify analog fields. The radar plot is a simple illustration that allows extending 2-D representations, and it is a first approach to multidimensional representations.

To illustrate how this comparison works for screening purposes, let us look at Figures 4.3 and 4.4, which show examples of radar plots that

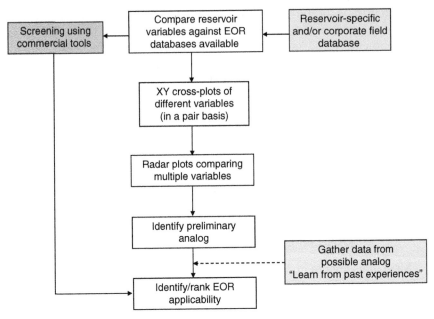

FIGURE 4.1   Conventional screening workflow.

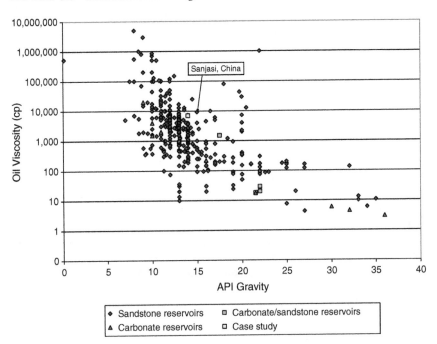

FIGURE 4.2   Example of an XY cross-plot comparing the oil gravity and the viscosity of the field under study (*square*) to the international steam injection (pilot and/or full-field) projects. In the example, a possible analog is shown.

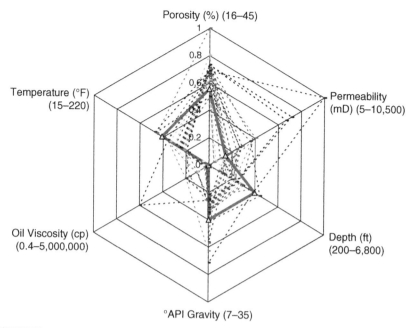

FIGURE 4.3 A radar plot comparing six reservoir variables of the field under study (*solid line*) with some international steam injection (pilot and/or full-field) projects. Here, the reservoir under evaluation lies within several projects with similar properties and so preliminarily passes steam injection applicability. (Triangles denote the data points.)

contain data from a given field against international steam injection and polymer flooding projects. Figure 4.3 shows an example where applicability is presumed, given that all of the reservoir properties evaluated lie within the range of parameters of known steam injection projects at either a pilot or a commercial scale. This stage of decision making in EOR does not take into account resource constraints. For the case in point, sources of gas and/or water availability to generate steam may or may not be part of the analysis (framing is necessary to decide that).

On the other hand, Figure 4.4 illustrates an example where polymer flooding is judged not applicable to the field being evaluated only on the basis of experience. This particular field exhibits much higher viscosities than do projects documented in the literature, which can often be used to consider discarding the applicability of polymer flooding. In this example, polymer flooding might not be appropriate because of the high polymer concentrations or high molecular weight required to reach the desired mobility control. The high viscosity of the polymer solution is expected to drastically reduce well injectivity, given the net pay and average reservoir permeability (flow capacity, kh) of this reservoir. Although the injection rate could be attained by increasing the number

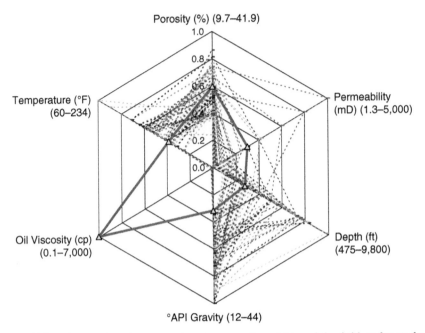

FIGURE 4.4 A radar plot comparing six reservoir variables of the field under study (*solid line*) with international polymer flooding (pilot and/or full-field) projects. Here, the reservoir under evaluation shows that oil viscosities are much higher than have been reported in project literature, thus eliminating the applicability of polymer flooding.

of injectors, the additional costs might make the project economically unattractive. This preliminary conclusion can be supported without the necessity of economic evaluations by using the analyst's intuition and/ or experience.

In the case of commercially unproven EOR processes (e.g., THAI, VAPEX), this screening phase helps you to analyze and estimate the technical feasibility of recovery methods based on more theoretical and engineering judgments. The notions that are conveyed by comparison with abundant field cases are the level of risk and bias containment, so "wisdom" does not become an excessive cognitive bias. Conventional screening is complemented by the use of screening options in commercial analytical tools to expand the evaluation and therefore further validate applicability (feasibility) of the most practical recovery process in the field under evaluation. Analytical screening is also based on the comparison of the reservoir properties of the field under evaluation with the property intervals of known IOR/EOR projects existing in each commercial tool.

Two known procedures are used to expand on the screening process. The first procedure is based on go–no go criteria considered in PRIze

(ARC, 2006; PRI, 1995), while the second one (SWORD) uses fuzzy logic to generate scores for ranking, based on the triangular distribution of comfort intervals (IRIS, 2007). Since the "bias" (expertise bias) differs in the two screening procedures, these additional approaches provide a comprehensive evaluation of the property/field of interest.

Although this first step of the proposed methodology represents a quick and useful approach, it can often be misleading for those who are unfamiliar with EOR technologies. For example, when evaluating the applicability of air injection in heavy-oil reservoirs (in situ combustion, or ISC) and/or in medium- and light-crude oil reservoirs (high-pressure air injection, or HPAI), engineers can get confused by different screening criteria that make it difficult to apply this recovery process in a particular field.

Table 4.1 shows an example of different screening criteria documented in the literature (Green and Willhite, 1998; Taber et al., 1997a; Turta and Singhal, 2001) that are included in PRIze criteria (ARC, 2006; PRI, 1995). Theoretically, if engineers (analysts) know the brackets of the screening criteria (see the table) for a particular reservoir, they should be able to identify the applicability of air injection. Assuming that the reservoir under evaluation is Cedar Hill, a dolomitic reservoir in North Dakota (also shown in Table 4.1), the applicability of air injection could be a challenging task.

Although it is possible for engineers to assume that air injection (HPAI in this case) is technically feasible, it is also possible that air injection can be ruled out given the low permeability values, net thickness, and transmissibility of Cedar Hills compared with the different screening criteria shown in Table 4.1. However, it is important to note that the Cedar Hills air injection project is one of the 11 (full-field and pilot) projects going on in Montana, North Dakota, and South Dakota (Gutiérrez et al., 2008; Manrique et al., 2007; Moritis, 2008; Watts et al., 1997). Nevertheless, details about the projects have been reported in the literature. The Cedar Hills example clearly shows that the applicability of EOR processes may not necessarily be ruled out if some of the criteria proposed by different experts or incorporated in commercial tools are not met.

Conventional screening of chemical EOR processes, such as alkali-surfactant-polymer (ASP) or surfactant-polymer (SP), is also a challenging task for engineers without EOR experience given the advances in the technology during the last two decades. Most of the micellar-polymer (MP) floods developed during the 1970s and 1980s were based on petroleum sulfonate surfactants (Aldrich et al., 1984; Ferrell et al., 1980; Heffern et al., 1982; Taggart and Russell, 1981). These surfactants were developed through the sulfonation of refinery distillate streams of a wide molecular weight range to produce the desired sulfonates or combinations thereof for EOR applications.

TABLE 4.1 Screening Criteria for Air Injection Processes

Variable	Screening Criteria for ISC and HPAI					Example: Cedar Hills, ND
	HTO-IAF	HTO-MAF	Taber et al. (1997)	EOR SPE Textbook	PRIze	
Formation Type	Sandstone / Carbonate	Sandstone / Carbonate	Sandstone	Sandstone	Sandstone	Dolomite
Depth, feet	>492	>6,562	<11,500	<11,500	492–5,900	8,500–9,000
Oil gravity, °API			10–27	>10		33
Oil viscosity (μ), cp	>10	>10	<500	<5,000	2–5,000	2
Oil density, kg/m^3		<850			>800	
Porosity (ϕ), %	>20	>15	High	High	>18	18
Average perm. (k), mD	>100	>10	>50	>50	>50	3–10
Pay thickness (h), feet	>9.8	>9.8	>10	>10	>9.8	9
Temperature, °F		>176	>100	>100		215–225
Minimum oil sat. (So)		>30%	>50 PV	>50		>50
Oil content, ϕSo	>0.065				>0.065	>0.09
Transmissibility, Kh/μ, md*ft/cp	>52.5		>20	>20	>52.5	13–50
Fractures	No	No			No	
Gas cap	No	No			Local	
Bottom water	No	No			Local	
Pressure		Pi > N$_2$ MMP				
Oil composition			Some asphaltic	Some asphaltic		

HTO = high-temperature oxidation; LTO = low-temperature oxidation; IAF = immiscible air flooding; MAF = miscible air flooding
Source: Compiled from Green and Willhite (1998) and Turta and Singhal (2001).

Despite a number of technical successes documented in the literature (Gogarty, 1978; Lowry et al., 1986), high concentrations and costs of surfactants and cosurfactants combined with low oil prices in the mid-1980s limited the use of micellar-polymer flooding. An additional advantage rarely documented in the literature is the long-term viability of producing large volumes of surfactants within the required specifications for EOR applications. In other words, as refinery feedstocks change with time (i.e., production decline of specific crude oils replaced by other crude oil types from different sources), the probability of generating the required petroleum sulfonates for a chemical EOR flood is much lower, increasing the risk for unsuccessful projects.

However, in the late 1980s, the old MP technology was replaced with synthetic surfactants that are highly tolerant to high salinities and temperatures are and applicable at very low concentrations. In other cases, the use of alkaline solutions to replace some of the high-cost surfactants used in micellar-polymer floods was also implemented, especially in low-salinity and divalent cation (hardness) content to avoid or reduce the precipitation of the alkaline solution.

To demonstrate the difficulty of conventional screening of chemical EOR processes, Table 4.2 shows a screening example of SP and ASP for a particular reservoir in Texas, using conventional methods that are commercially available. Screening results show that the temperature and water hardness for the reservoir under evaluation are higher than the recommended criteria for SP and ASP. It is likely that an engineer with no experience in the area might discard both recovery processes or consider the identification of possible water sources for injection and/or a water-softening strategy to justify using these EOR methods. However, even at this early stage of the evaluation, a review of water softening costs and the availability of additional water sources for injection and its costs (i.e., the need to drill wells to produce the water) is recommended.

On the other hand—based on advances in surfactant technologies— several field applications (Hernández, Alvarez, et al., 2002; Manrique et al., 2007; Pitts et al., 2006; Pratap and Gauma, 2004) and laboratory studies (Aoudia et al., 2007; Levitt et al., 2009; Manrique et al., 2000; Pandey et al., 2008) report SP or ASP evaluations at higher temperatures (up to 90 °C), water hardness (> 350 ppm), and salinities (up to 200 g/l). Although prior experience does not guarantee the technical or economic success of chemical EOR processes at present, it is important to develop a comprehensive review of data sources to complement and justify EOR screening results.

Figure 4.5 shows some examples of surfactant types that can be used under different reservoir salinities and temperatures. If the evaluator–engineer who is performing the EOR screening studies is part of or has

TABLE 4.2   Screening Criteria for Surfactant-Polymer and Alkali-Surfactant-Polymer

Screening Parameters(a)	Reservoir under Evaluation in Texas	Screening Criteria for (a) Surfactant-Polymer	Screening Criteria for (a) Alkali-Surfactant-Polymer
Formation Type	Sandstone		Sandstone
Reservoir temperature, °C	75	<70	<70
Oil density at surface, kg/m^3	925		>850
Live oil viscosity at P_b, mpa*s	4.2	<150	<150
Horizontal permeability, md	300–1,500	>50	>50
Active water drive (yes/no)	No	No	No
Bottom water: none/local/ margins/extensive	Local	≤local	≤ local
Gas cap: none/local/ margins/extensive	None	≤local	≤ local
Clay content: none/low/high	Low	≤low	≤low
Water hardness, ppm	1,953	<1,000	<20
Water salinity, ppm	8,567	<35,000	<35,000
Current oil saturation, fraction	0.542	0.350	0.350

Note: Values in bold do not meet screening criteria for both chemical EOR processes SP and ASP.
(a) Source: From PRIze analytical simulator.

access to a large organization or university with laboratory capabilities, the consolidation of phase behavior and surfactant stability tests, among others, in simple matrices as shown in Figure 4.5 or in databases will continuously improve screening studies, making them more robust to support proper decision-making processes.

Although conventional commercially available screening tools are still valuable in estimating EOR applicability, the latter clearly demonstrates the necessity in update screening methods that incorporate recent advances in different EOR technologies. Therefore, one must not blindly take screening outcomes at face value. EOR screening should evolve to account for advances in technologies (e.g., new chemicals) as well as field experiences from majors to independents that might not have been documented in the literature but can be found on the Internet. In the meantime, it is a good idea to develop conventional screening and incorporate as many methods and sources of information as possible, mitigating the chance of "bias."

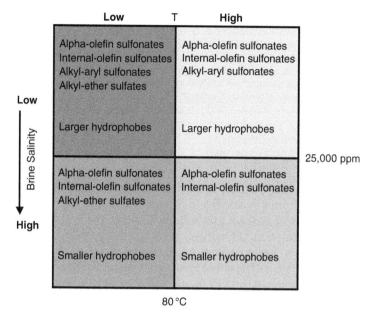

FIGURE 4.5 Types of surfactants that can be considered for chemical EOR (SP or ASP) at different reservoir salinities and temperatures. *Source: Courtesy of TIORCO LLC.*

4.3 GEOLOGIC SCREENING

Conventional EOR screening criteria have been widely used to evaluate numerous reservoirs before performing any detailed evaluations. Conventional criteria often include coarse geologic indicators—that is, whether it is a clastic (sand) or a nonclastic (carbonate) reservoir—or perhaps a coefficient to serve as a heterogeneity indicator (e.g., the Dykstra-Parsons coefficient). These representations may suffice at early stages of screening, but there are controls on EOR processes that require more detailed information on the reservoir geology to be assessed. This is the purpose of geologic screening, and we will focus on the critical geologic aspects. Despite their importance, geologic screening criteria ("predictive geology") have not been used as frequently as other forms of screening.

Geologic analogs for the estimation of petroleum reserves, a commonly used tool in the oil and gas industry (Hodgin and Harrel, 2006; SEC, 2008; SPE, 2007), are meant to be used in further screening. To understand the concept of analogous reservoirs, let us quote the definition (S-X 210.4-10(a)(2)) that was recently offered by the U.S. Securities and Exchange Commission (2008):

Analogous reservoirs, as used in resources assessments, have similar rock and fluid properties, reservoir conditions (depth, temperature, and pressure), and drive mechanisms, but are typically at a more advanced stage of development than the reservoir of interest and thus may provide concepts to assist in the interpretation of more limited data and estimation of recovery. When used to support proved reserves, an "analogous reservoir" refers to a reservoir that shares the following characteristics with the reservoir of interest:

(i) Same geological formation (but not necessarily in pressure communication with the reservoir of interest);
(ii) Same environment of deposition;
(iii) Similar geological structure; and
(iv) Same drive mechanism.

It is not always easy to meet all of the SEC or SPE criteria to prove the technical and economic feasibility ("Reserves commerciality") of a particular EOR process based on previous field experiences similar to a reservoir under evaluation. However, geologic and reservoir analogies definitively serve to build estimators of resources when direct data are limited or not available. These analogies contribute to reducing potential uncertainties when evaluating EOR applicability or incremental recovery factors in a particular reservoir under evaluation. A good example of an SEC reservoir geologic analogy is the HPAI projects reported in light-oil reservoirs in Montana, North Dakota, and South Dakota (Moritis, 2008). Moritis reports up to 11 active HPAI projects in carbonate formations in the United States.

HPAI evaluations started in the late 1970s, with a few new projects during the 1980s. However, more than half of the projects started in the 2000s. In Section 4.1, we demonstrated that the applicability of HPAI could be ruled out for Cedar Hills on the basis of the low permeabilities, net thickness, and transmissibility (see Table 4.1) using conventional screening criteria. However, it is important to indicate that all 11 HPAI projects reported in Montana, North Dakota, and South Dakota, were developed in the same geologic formation (Red River A, B, and C).

In addition, most of these reservoirs exhibit similar geologic and reservoir properties based on information available in the U.S. DOE Toris database that meets SEC and SPE reservoir analog definition criteria. Table 4.3 shows basic geologic and reservoir properties that the Red River formation has in common with several fields in Montana, North Dakota, and South Dakota (Toris, 1995).

Geologic characteristics, such as trap type, depositional environment, geologic age, lithology, type of structure, and digenesis, are used to establish a comparison basis between a field under evaluation and EOR projects recorded in a database or information documented in the literature. Several studies have demonstrated the use of reservoir geologic analogy to determine the technical feasibility or applicability of EOR in a particular field. However, most studies have been documented

TABLE 4.3   Red River Formation Common Reservoir and Geologic Properties

Lithology	Carbonate
Heterogeneity	Diagenetic overprint
Trap type	Mostly combination (stratigraphic/structural)
Structure	Unstructured
Diagenesis	Dolomitization
Depositional system	Peritidal/intertidal
Geologic age	Silurian/Ordovician
Depth, ft	8,300–12,700
Temperature, °F	193–288
Net pay, ft	16–44
Porosity, %	6–15
Permeability, md	2–37

Source: Compiled from Torris, 1995.

mostly for sandstone formations (Caers et al., 2000; Cokinos et al., 2004; Gachuz-Muro and Sellami, 2009; Henson, 2001; Knapp and Yang, 1999; Manrique and Pereira, 2007) and significantly less documented for carbonate formations (Allan and Sun, 2003).

For sandstone reservoirs, this analysis can be augmented by using the matrix of depositional environment versus lateral and vertical heterogeneities (Henson, 2001; Henson, Todd, and Corbett, 2002; Tyler and Finley, 1991). The cited studies show the relationship between a sandstone reservoir's architecture and both conventional and EOR strategies recovery efficiencies. This type of analysis consists of a matrix containing the depositional systems characteristics, mainly the lateral and vertical heterogeneity indicators. The classification, although generally subjective because of the lack of geologic information and/or differences in the geologic interpretation, can still guide the EOR decision-making process using field experience.

Figure 4.6 shows this classification scheme for 200 steam flooding projects in sandstone reservoirs that were plotted in the Tyler and Finley (1991) heterogeneity matrix. Of the 200 steam flooding projects described, 185 and 15 were reported by the operators as successful and failed, respectively.

As can be seen in Figure 4.6, most of the steam floods have been developed in depositional systems with moderate vertical heterogeneity (VH) and moderate to high lateral heterogeneity (LH). For example, to

		Lateral Heterogeneity		
		Low	Moderate	High
Vertical Heterogeneity	Low	Wave-dominated delta Barrier core Barrier shore face Sand-rich strand plain (9)	Delta-front mouth bars Proximal delta front (accretionary) Tidal deposits Mud-rich strand plain (7) / [3]	Meander belts* Fluvially dominated delta* Back Barrier* (0)
	Moderate	Eolian Wave modified delta (distal) (9) / [2]	Shelf barriers Alluvial fans Fan delta Lacustrine delta Distal delta front (83) / [9]	Braided stream Tide-dominated delta (52)
	High	Basin-flooring turbidites (19)	Coarse-grained meander belt Braid delta (2)	Back barrier** Fluvially dominated delta** Fine-grained meander belt** Submarine fans** (4)

* Single units;   **Stacked systems

FIGURE 4.6   Steam flooding as a function of the depositional system (heterogeneity matrix) of sandstone reservoir proposed by Tyler and Finley. The numbers in parentheses represent successful; numbers in brackets represent unsuccessful projects.

evaluate the technical feasibility of a steam flood in a reservoir with a depositional system of tidal deposits, at least ten projects have been reported in depositional systems with low VH and moderate LH (see Figure 4.6), out of which seven were reported as successful.

This in no way means that the project under consideration is not technically feasible because a comparison is made with projects developed in different depositional systems. However, if they incorporate EOR conventional screening methods and additional information into the analysis, such as the reservoir dip (21°) and the presence of an active aquifer that may lead to steam segregation and important heat losses, respectively, the asset team and the decision maker may have enough information to rank this project against other investment opportunities at this early stage of evaluation.

This example demonstrates how this type of analysis can complement the risk management strategies of a particular oil company or investor. Even though it has been incorporated into the evaluation of steam-assisted gravity drainage (SAGD) projects in Canada (Manrique and Pereira, 2007), we recognize that the Tyler and Finley heterogeneity matrix is not necessarily representative of the proper depositional systems classification of Canadian oil sands. However, we believe that this approach is still valuable for property evaluation and acquisitions.

If the dimensions of sand bodies or genetic units (length, thickness, and width) and current or proposed well length and spacing are known, horizontal and vertical heterogeneities indexes can be estimated through simple equations. In this approach, the lateral heterogeneity index (LHI) is estimated for different "genetic unit mean lengths" (GUML) and completed well intervals, while the vertical heterogeneity index (VHI) is estimated for different "genetic unit mean thicknesses" (GUMT).

The results are dimensionless, and the minus sign is there to give a positive heterogeneity index that increases with increasing heterogeneity. Heterogeneity (2-D) indexes for vertical wells can be calculated using the following equations (Henson, Todd, and Corbett, 2002):

$$\text{Lateral Heterogeneity Index (LHI)} = -\log\frac{\text{Genetic Unit Mean Length}}{\text{Well Spacing}}$$

and

$$\text{Vertical Heterogeneity Index (VHI)} = -\log\frac{\text{Genetic Unit Mean Thickness}}{\text{Gross Pay Thickness}}$$

Heterogeneity (2-D) indexes between two horizontal wells where the flow between them is vertical (i.e., SAGD well pairs) can be estimated from the following equations:

$$\text{Lateral Heterogeneity Index (LHI)} = -\log\frac{\text{Genetic Unit Mean Length}}{\text{Completed Well Interval}}$$

and

$$\text{Vertical Heterogeneity Index (VHI)} = -\log\frac{\text{Genetic Unit Mean Thickness}}{\text{Vertical Well Spacing}}$$

Additionally, the data that is required to estimate these heterogeneity indexes for a particular reservoir can be obtained from core and log data, well test analysis, and the like. If the data are not available, information from reservoir analog (Hodgin and Harrel, 2006; SEC, 2008; SPE 2007) could be used as a preliminary assessment. The methodology proposed by Henson, Todd, and Corbett (2002) can be used to evaluate the impact of reservoir heterogeneities on EOR processes. The following example will demonstrate how the proposed methodology can be used to evaluate the impact of reservoir heterogeneities on SAGD projects.

Several authors have reported how reservoir architecture and heterogeneity influence the efficiency of SAGD processes (Fattahi et al., 2004; McCormack, 2001; Putnam and Christensen, 2004; Rottenfusser and

Ranger, 2004). The most critical factor for SAGD projects is the ability of the steam chamber to confine the injected fluids, which facilitates the recovery of mobilized oil. Therefore, evaluating reservoir heterogeneities that affect the development of the steam chamber, and consequently the ultimate recovery of the oil, is a key activity that reduces some of the risk associated with this recovery process. Figure 4.7 shows how vertical and lateral heterogeneities may affect the performance of SAGD projects.

In summary, the heterogeneity matrix that is shown in Figure 4.6 can be interpreted as follows:

- The lower the heterogeneity indexes (negative values), the higher the probability that a SAGD well pair will be in the same sand body (sand channel). Low indexes provide good opportunities for developing stable steam chambers and thus SAGD projects (environments of low to moderate lateral and vertical heterogeneities).
- As the heterogeneity indexes increase, the probability that SAGD well pairs will be in different sand bodies also increases. This condition may affect the development of the steam chamber and thus the ultimate recovery factor (environments of moderate to high lateral and vertical heterogeneities).

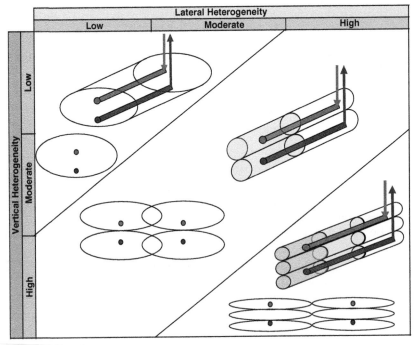

FIGURE 4.7 The effect of vertical and lateral heterogeneities on recovery efficiencies of SAGD projects.

To better explain how these heterogeneity matrices can be used for different evaluations, Figure 4.7 illustrates a simplified example of the influence of vertical and lateral heterogeneities on horizontal extension and vertical separation of SAGD well pairs. The heterogeneity matrix shown in Figure 4.8 represents an example for an SAGD feasibility study in the McMurray formation, Canada.

This heterogeneity matrix is for an average GUML of 1,100 m and different GUMT (15, 30, and 45 m), horizontal section length (700–1,000 m), and vertical separation (3, 5, and 7 m) of SAGD well pairs. In summary, Figure 4.8 can be interpreted as follows:

- The lower the heterogeneity indexes (the negative values at the top left of the plot), the higher the probability that a SAGD well pair will be in the same sand body (sand channel) and thus the higher the probability of increased recovery factors.
- The thicker the pay (GUMT) in the McMurray formation, the lower the heterogeneities showing a good potential to develop successful SAGD projects. (See the legend in Figure 4.8 for GUMT = 45 m.) In this analysis, evaluating whether GUMT consists of continuous pay is not possible. Therefore, this analysis must be complemented with basic log interpretations, including the Dykstra-Parsons calculations, if proper data are available. It is important to note that the presence of the low-permeability layers interbedded within the pay thickness may negatively impact the development of steam chambers, reducing the potential to increase oil recoveries in SAGD projects.

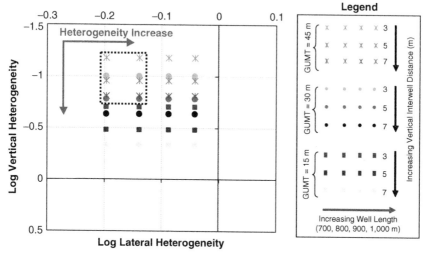

FIGURE 4.8 Heterogeneity matrix (*left*) showing the influence of vertical and lateral heterogeneities on horizontal extension and vertical separation in SAGD well pairs.

- For a constant GUML of 1,100 m, the larger the horizontal section of SAGD well pairs, the higher the lateral heterogeneity. (In Figure 4.8, the horizontal section length increases from left to right for a constant VHI or log heterogeneity index.) However, drilling SAGD wells with longer horizontal sections will depend not only on reservoir heterogeneities but also on parameters such as heat loss and a reduction in steam quality. Consequently, it will affect the critical operational parameters of SAGD projects (Herrera and Mendoza, 2001; Jimenez, 2008; Kaura and Sierra, 2008; Krawchuk et al., 2006; Lerat et al., 2009).
- The higher the vertical separation of SAGD well pairs, the higher the LVH. As an example, for a GUMT of 45 m and a constant LHI of −0.196, the VHI increases from −1.176 to −0.808 when SAGD well pairs are separated from 3 to 7 m, respectively. Vertical separation of SAGD well pairs of approximately 5 m are common in Canadian SAGD pilot and commercial projects (Albahlani and Babadagli, 2008; Good et al., 1994; Manrique and Pereira, 2007; Parmar et al., 2009).
- In the example SAGD, well pairs with lateral extensions of 700 to 800 m and vertical separations of 5 m are considered at lower risk for GUMT (thicknesses) of 30 m and more (*dashed square* in the figure).

Heterogeneity matrices can also be generated to estimate well spacing based on reservoir heterogeneities. Well spacing of SAGD well pairs is a critical parameter for the economic success of field projects (i.e., steam–oil ratios, steam chamber interference between well pairs, CAPEX, and OPEX, among others) (Akram 2008a, 2008b; Jimenez, 2008; Manrique and Pereira, 2007; Singhal et al., 1998; Vanegas et al., 2009). Although it must be recognized that well spacing of SAGD well pairs is influenced by reservoir properties and operational conditions, spacing between 70 and 100 m has generally been proposed.

In cases where representative geological models and/or interpretations are not available, 2-D and 3-D heterogeneity index analyses have been shown to be more robust when combined with Dykstra-Parsons (DP) coefficients calculated from well logs (petrophysical analysis) and core permeability data. However, if permeability data from well logs and cores are unavailable, the identification of potential vertical low-permeability or no-flow barriers (i.e., shales or siltstones) through basic review of well logs is also useful. Vertical communication within the pay zone is generally important for different EOR processes, especially those for SAGD and gravity drainage, such as gas injection and double displacement.

Because one of the main objectives of this book is to propose fast screening methods for EOR projects, the authors consider the use of DP one such option for analytical and numerical simulations as a preliminary approach and until geological and upscaled numerical models ("hopefully") become available, as well as the required data for proper numerical simulation studies.

Permeability variation represents a good indicator of reservoir heterogeneity. The DP coefficient has been the traditional representation of reservoir heterogeneity based on permeability variation. The DP coefficients vary from 0 for homogeneous reservoirs (or very low-permeability variations) to 1 for highly heterogeneous reservoirs (or very high-permeability variations). As was mentioned before, petroleum or reservoir engineers may have to develop preliminary assessments of the performance of a particular reservoir under a particular recovery process when geologic models are unavailable or detailed analysis is not justified with the information available. In such cases, using a simplistic approach to represent the level of heterogeneity based on DP coefficients is very useful in EOR screening studies, especially when analyzing a large portfolio of reservoirs.

As we saw before, DP coefficients can be calculated from well logs (petrophysical analysis) and/or core permeability data. The estimated DP coefficients can be determined in several different ways (i.e., graphically or numerically), and this has been documented by Green and Willhite (1998) and Jensen and Currie (1990), among others. To determine the permeability distribution of a particular set of data, it is necessary to calculate the cumulative probability distribution of the raw permeability data. From the data shown in Table 4.4, the permeability variation (DP coefficient) can be calculated using the following equation:

$$DP = \frac{k_{50} - k_{84.1}}{k_{50}}$$

where k_{50} and $k_{84.1}$ are the permeability data for the 50 and 84.1 cumulative probabilities, respectively.

In this example, a set of 33 permeability data points were recorded from log information. The cumulative probability was estimated to fit the data to a log-normal permeability distribution, assuming that it could be approximated to a straight line around the data's midpoint (50th percentile). Cumulative probability (the third column in Table 4.4) represents the fraction of the samples with permeabilities greater than a particular sample. In Table 4.4, the screened cells represent the permeability data closest to Pi(k); the 50 and 84.1 cumulative probabilities obtained a DP coefficient of 0.672.

DP coefficients have also been used extensively to generate full-field maps as a quick quality-control procedure during petrophysical studies (e.g., the impact of petrophysical cutoffs on reservoir heterogeneity) as part of detailed reservoir ("integrated") engineering studies. DP maps combined with other reservoir properties (e.g., net pay and fluid saturation) also have been used in full-field analytical simulations to evaluate

TABLE 4.4 Cumulative Probability for Permeability Data

Records	Permeability (k), mD	Cumulative Probability
1	563.94	
2	475.23	0.030
3	393.70	0.061
4	353.33	0.091
5	341.95	0.121
6	297.15	0.152
7	270.80	0.182
8	261.25	0.212
9	254.32	0.242
10	254.26	0.273
11	248.01	0.303
12	226.59	0.333
13	216.03	0.364
14	169.55	0.394
15	169.38	0.424
16	130.82	0.455
17	110.15	0.485
18	108.70	0.515
19	103.40	0.545
20	101.95	0.576
21	96.11	0.606
22	91.81	0.636
23	88.37	0.667
24	86.52	0.697
25	85.17	0.727
26	84.49	0.758
27	83.12	0.788
28	36.13	0.818
29	27.22	0.848
30	19.92	0.879
31	16.07	0.909
32	12.06	0.939
33	7.92	0.970

RDP under EOR processes (Alvarado et al., 2003). For more details, please refer to Chapter 3 of this book, Simulations and Simulation Options.

Most, if not all, of the analytical simulators include the capability to predict waterflooding and/or EOR performance using the DP approach. However, this approach also can be used in numerical simulators, as discussed in Chapter 3 on field cases. DP grids can be generated from a given permeability range and average permeability (i.e., the More-Roxar simulator), as shown in Figure 4.9.

This simple simulation grid is extremely helpful for estimating and ranking the potentials of different EOR processes at the preliminary stages of evaluation and when geological or simulation models are not available. Such grids can also be generated with the help of special geomodeling software, using different depositional environments, as described by Henson (2001) to estimate the impact of depositional environments on the IOR/ EOR potential in a particular reservoir or portfolio of reservoirs.

A few closing remarks for this section are in order. Geologic screening is not a replacement for simulations, but it may be used as an early strategy and a very powerful tool to examine EOR methods that are highly controlled by reservoir heterogeneity (and many are) and continuity. SAGD and water-alternating-gas injection (WAG) are just two of the frequently encountered examples. This type of screening allows practitioners to scrutinize critical variables and conditions in a reservoir or field that would require a more detailed analysis, with the consequent

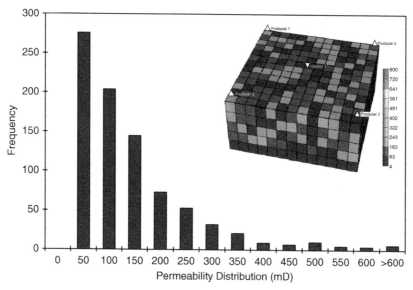

FIGURE 4.9 A Dykstra-Parsons simulation grid based on a given permeability distribution and average permeability of 140 mD.

investment of time and resources that may not be necessary because of the limited applicability of an EOR process.

Geologists may argue about the incomplete or inadequate representation in this type of screening, and they may very well be correct, but then this begs the question: Do we ever have a complete geologic representation without access to detailed integrated studies? The answer is likely to be no in most cases. Therefore, the idea of using a combination of screening methods will complement the analysis, making it possible to perform quick evaluations before long and expensive studies are undertaken. Now we will examine the overall strategy of estimating the technical and economic feasibility of EOR processes.

4.4 ADVANCED EOR SCREENING

In the proposed EOR screening methodology, advanced EOR screening refers to more robust data mining strategies and artificial intelligence techniques that can lead to better screening criteria by considering simultaneous combinations of more than two reservoir and fluid properties. Artificial intelligence, specifically neural networks, fuzzy logic, and expert systems, have been widely proposed and used for supporting multiple applications in oil and gas operations (Abdulraheem et al., 2009; Alegre et al., 1993; Ali, 1994; Allain and Houze, 1992; Balch et al., 2000; Hamada and Elshafei, 2009; Hutchin et al., 1996; Mohaghegh, 2000; Peden and Tovar, 1991; Weiss et al., 2000).

Different artificial intelligence techniques have been used to develop screening and a selection of enhanced oil recovery methods (Alvarado et al., 2006; Chung et al., 1995; Elem and Elmtalab, 1993; Gharbi, 2000; Guerillot, 1988; Ibatullin et al., 2002; IRIS, 2007; Jian and Wenfen, 2006; Shokir et al., 2002; Surguchev and Li, 2000).

To overcome some of the limitations of conventional look-up tables that list screening ("go" and "no-go") criteria, the proposed advanced screening methodology resorts to artificial intelligence (AI) techniques. The development of this strategy has been well documented in the literature (Alvarado et al., 2002; Hernández, Liscano, et al., 2002; Manrique et al., 2003). The method is based on data mining of a database of international projects, for which the applicability and success of recovery processes were collated along with reservoir and fluid data. The data mining process yields a new strategy for screening oil recovery methods (IOR and EOR). It is based first on space reduction techniques to simplify the representation of international experience on oil recovery methods, represented in a collated database of reservoirs and projected as 2-D cluster maps ("expert maps").

A statistical analysis of the data shows how important the variables are in terms of how they influence reservoir clusters. A small number of these variables, which represent average values for each reservoir, are used to rank oil recovery methods, and then a knowledge discovery database (KDD) strategy is used to extract knowledge for the selection of oil recovery methods for their application in reservoirs (Mitchell, 1997). The primary goal of this screening approach is to develop an artificial intelligence framework based on KDD, using a reduced set of characteristic reservoir variables. The end result of data processing is the identification of possible reservoir clusters or typologies. The combination of the reduced-space representation with the KDD approach opens up a different way of screening IOR/EOR methods (Alvarado et al., 2002; Hernández, Liscano, et al., 2002; Ranson et al., 2001; Ranson et al., 2002).

Before we proceed with our discussion of this advanced screening approach, it is worth mentioning that this is the only step of the methodology developed by the authors to evaluate EOR potential that is not publicly available. All of the steps of the proposed methodology for EOR feasibility studies can be developed using different tools and do not depend on a particular commercial software platform. Of course, the availability of data on EOR projects represents an advantage. However, most of the information is in the public domain (e.g., *Oil & Gas* EOR–production surveys and U.S. DOE databases) or is available commercially and can be processed through advanced artificial intelligence techniques that are already incorporated in commercial software tools. Therefore, advanced EOR screening approaches can be extended and adapted according to the user's specific needs.

The database used to generate the developed advanced screening approach contains information from oil recovery projects carried out all over the world, which allows one to compare the reservoir of interest with projects executed in more than one country or continent. Most of these projects have been completed in the United States and Canada; the remaining projects have been carried out in many other countries and on several continents, including Asia, Europe, and Latin America. The database includes a list of more than 20 reservoir and fluid variables, although some records are missing information for some of them. More than 20 variables were initially considered, but in practice having a large number of records from public sources for the analysis with valid data for these 20 variables is almost impossible. A careful variable selection process was required using all of the available information.

Different map realizations were performed, and it was shown that six variables were sufficient to generate the maps. The selection process was based on the importance of those six variables to form well-defined clusters in the map; also, some reduction of the redundant information was done using correlation analysis (e.g., temperature vs. depth). Once the database was created and the main variables selected, the next step was to create the compressed representation of field experiences.

After realizing the space reduction, and despite the subjective nature of the resulting 2-D or 3-D data representations, it was possible to visually identify reservoir clusters, which heuristically are named reservoir types, typologies, or analog (Figure 4.10). A point on the 2-D maps represents each reservoir; however, if the variability of reservoir parameters is considered, then a neighborhood (a cloud of points) is shown. Groups of points forming clusters are what we have defined as a reservoir typology.

In this sense, it is observed that certain types of reservoirs tend to group in specific areas of the maps, and these reservoirs have in common the recovery methods that have been applied to them. So, through a direct inspection of the map, it is possible to screen reservoirs and applicable recovery methods. The proposed approach allows us to carry out fast and clean screening of the recovery methods based on the "reservoir pseudo typology" (Alvarado et al., 2002; Hernández, Liscano, et al., 2002; Ranson et al., 2002).

Figure 4.10 shows a 2-D representation (expert map) based on the simultaneous projection of a reduced set of reservoir variables—namely, reservoir depth, current reservoir pressure, porosity, permeability, crude oil gravity (API), and viscosity. Expert maps show that the 2-D representations of reservoir clusters have in common the types of EOR projects implemented. Multidimensional projections on the 2-D plots lead to the simultaneous comparison of multiple variables and, more important, a convenient clustering of reservoir types. It is in this way that statistics on recovery factors can be obtained, adding robustness and reducing the influence of "expert opinions" on the screening evaluation.

It is important to mention that from the resulting 2-D representations (see Figure 4.10) a complete set of polar axes are generated. Each axis represents the magnitude and behavior of each variable in the map. This information is very useful because it helps to explain why reservoir groups are mapped in a specific position. Using the axis and the behavior of each variable for locating a reservoir in the representation, a transformation matrix is developed so that it is possible to project new reservoirs (Ranson et al., 2002).

For example, at the top right-hand corner of the expert map, cluster 6 can be identified (see Figure 4.10). This cluster is characterized by deep, high-pressure/high-temperature (HP/HT) reservoirs that have miscible gas flooding (hydrocarbon and N_2) as the most common recovery process in this reservoir typology. On the other hand, on the opposite side of the expert map (left side of the 2-D projection in the figure), cluster 5 is dominated by shallow low-pressure and viscous oil reservoirs.

As expected, EOR thermal methods represent the most common of the recovery processes in this reservoir typology. Indeed, approximately

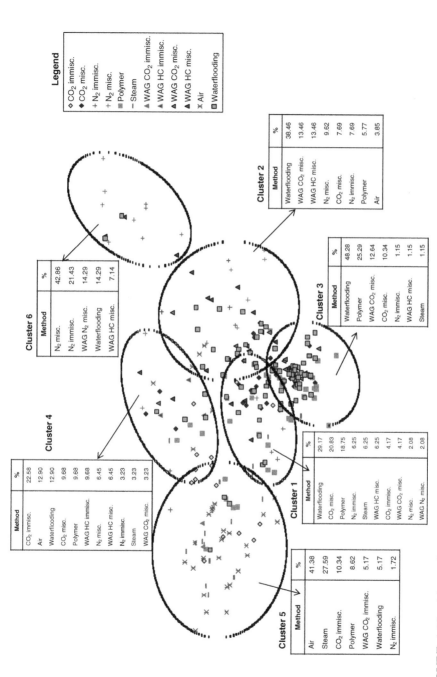

FIGURE 4.10 Advanced screening (expert) map. 2-D representation of multidimensional analysis of six reservoir variables: depth (TVD), current reservoir pressure, porosity, permeability, crude oil API gravity, and viscosity.

Legend

◇ CO₂ immisc.
◆ CO₂ misc.
+ N₂ immisc.
+ N₂ misc.
■ Polymer
− Steam
▲ WAG CO₂ immisc.
▲ WAG HC immisc.
▲ WAG CO₂ misc.
▲ WAG HC misc.
✕ Air
■ Waterflooding

Cluster 6

Method	%
N₂ misc.	42.86
N₂ immisc.	21.43
WAG N₂ misc.	14.29
Waterflooding	14.29
WAG HC misc.	7.14

Cluster 2

Method	%
Waterflooding	38.46
WAG CO₂ misc.	13.46
WAG HC misc.	13.46
N₂ misc.	9.62
CO₂ misc.	7.69
N₂ immisc.	7.69
Polymer	5.77
Air	3.85

Cluster 3

Method	%
Waterflooding	48.28
Polymer	25.29
WAG CO₂ misc.	12.64
CO₂ misc.	10.34
N₂ immisc.	1.15
WAG HC misc.	1.15
Steam	1.15

Cluster 4

Method	%
CO₂ immisc.	22.58
Air	12.90
Waterflooding	12.90
CO₂ misc.	9.68
Polymer	9.68
WAG HC immisc.	9.68
N₂ misc.	6.45
WAG HC misc.	6.45
N₂ immisc.	3.23
Steam	3.23
WAG CO₂ misc.	3.23

Cluster 1

Method	%
Waterflooding	29.17
CO₂ misc.	20.83
Polymer	18.75
N₂ immisc.	6.25
Steam	6.25
WAG HC misc.	6.25
CO₂ immisc.	4.17
WAG CO₂ misc.	4.17
N₂ misc.	2.08
WAG N₂ misc.	2.08

Cluster 5

Method	%
Air	41.38
Steam	27.59
CO₂ immisc.	10.34
Polymer	8.62
WAG CO₂ immisc.	5.17
Waterflooding	5.17
N₂ immisc.	1.72

70 percent of the typology's projects have been developed either by in situ combustion or steam injection. These examples briefly explain how each axis represents the magnitude and behavior of each variable or its combinations (e.g., sensitivity analysis) in the map.

The following sections present some examples of evaluations completed by the authors during the last few years.

4.4.1 Identifying Reservoir Analog

In this type of screening, "reservoir analog" or "closer analog" refers to a particular field that has common reservoir properties and a pilot test or large EOR project (ongoing or implemented) of interest to the field/reservoir under evaluation. In other words, two reservoirs (represented as data points in the expert map in Figure 4.10) closely located in the same "reservoir typology" or cluster usually exhibit similarities that contribute to validating the technical feasibility of a particular recovery process. This analysis can help to reduce potential uncertainties associated with the applicability of a particular recovery process in the reservoir under evaluation (e.g., WAG-CO_2 in low permeable and deep formations due to a potential lack of injectivity or EOR chemical flooding in high-salinity formations, among others).

Therefore, when we speak of the "reservoir analog" in this analysis, it must be recognized that it does not comply with the SEC definition of reservoir analogy described in Section 4.3. However, the proposed advanced screening can be adapted to meet SEC standards if data are available. One good example is the U.S. DOE Toris database, which reports key variables and their level of confidence (geologic age, geologic play, province, depositional environment, formation name, diagenetic overprint, and various other reservoir characteristics) that are required to define a reservoir analog.

To perform this advanced screening, data that was mined from roughly 450 successful EOR projects are compared with those of the reservoir(s) under evaluation. Figure 4.11 shows an example of an evaluation of the technical feasibility of enhanced oil recovery chemical flooding in a saline reservoir located in South America (Field A). The first step of this evaluation was to update the expert map shown in Figure 4.10 with several EOR chemical projects such as alkali-polymer (AP), surfactant-polymer (SP), and alkali-surfactant-polymer (ASP). In this way, the statistics on recovery processes can be updated, adding robustness and reducing "expert opinions" to the screening evaluation.

Once the expert map has been updated with projects of interest, the reservoir under evaluation generally undergoes conventional screening, including the use of analytical tools. From the conventional screening and from interaction with the reservoir and production engineers who

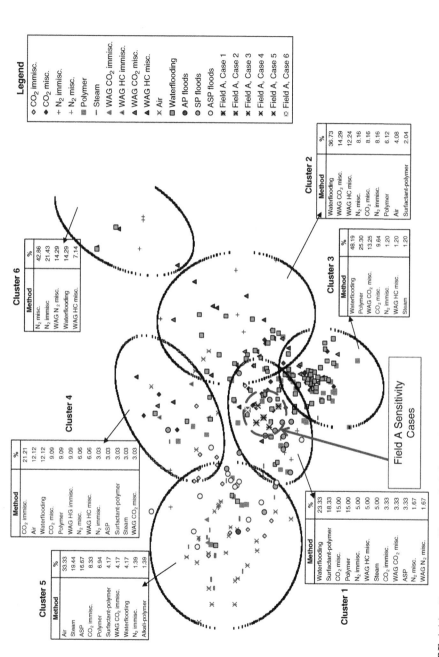

FIGURE 4.11 Expert map with extended chemical EOR (AP, SP, and ASP) projects to identify potential analog for a field (A) under evaluation.

are operating the field under evaluation, a range of variables can be agreed on to develop a sensitivity analysis as part of framing and defining the objectives of the analysis.

Table 4.5 summarizes input data used to perform a sensitivity analysis as part of this example. The main objective of this analysis considers the following cases:

- Case 1 is assumed to be the base case.
- Reservoir temperature and crude oil gravity are kept constant, except for Case 6, which is based on some discrepancies observed in different sources of information.
- Cases 1, 2, and 3 evaluate the permeability (e.g., uncertainties related to polymer injectivity) and porosity effects that are keeping the reservoir pressure, crude oil gravity, and viscosity constant.
- Cases 1, 4, and 5 evaluate the reservoir pressure and crude oil viscosity effects that are maintaining crude oil gravity, permeability, and porosity constant.

The results of the sensitivity analysis of Field A to identify potential "reservoir analog" or "closer analog," where EOR chemical flooding has been tested at field scale, are shown in Figure 4.11. All Field A sensitivity cases (cases 1 through 6) fall within cluster 1 of the expert map. It is important to notice that more than one-third of the projects ($\cong$36 percent) implemented in this reservoir typology have been SP, polymer flooding (P), and ASP. However, the information used to generate the expert maps in the figure can be handled at different levels of granularity, allowing a more detailed (close-up) analysis and adding more value to the decision-making process.

Figure 4.12 shows a zoom view of cluster 1 of the expert map shown in Figure 4.11. This representation corresponds to a different degree of

TABLE 4.5 Input Data for Performing a Sensitivity Analysis of Field A

Cases	Porosity (%)	Temp. (°F)	Pressure (psi)	Permeability (md)	Oil Gravity (°API)	Oil Viscosity (cp)
Case 1	18	120	1,000	100	34	2
Case 2	15	120	1,000	50	34	2
Case 3	18	120	1,000	200	34	2
Case 4	18	120	1,500	100	34	1
Case 5	18	120	500	100	34	3
Case 6	18	149	1,000	100	34	2

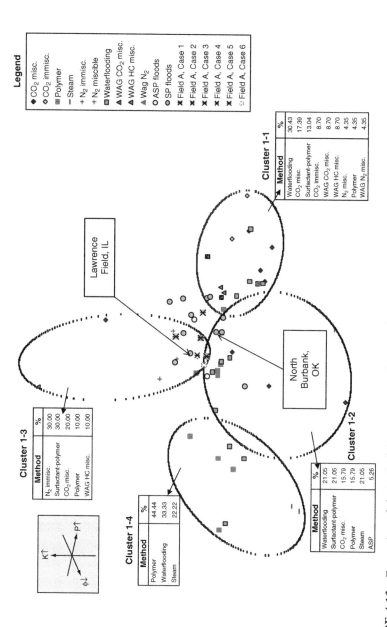

FIGURE 4.12 Zoom view of cluster 1 of the expert map shown in Figure 4.11 to identify potential analog for a field (A) under evaluation for EOR chemical flooding.

granularity where the new clusters result from the hidden structure inside cluster 3 of the initial (Figure 4.11) representation.

Figure 4.12 also shows (in the screened box at the top left of the figure) three of the six polar axes representing the magnitude of variables such as permeability, porosity, and reservoir pressure. For example, projects located on the right side of cluster 1-1 are reservoirs with the highest reservoir pressures of the group of projects of this reservoir typology (see figure). Again, this information is very useful when analyzing the sensitivity of dynamic properties like reservoir pressure and crude oil viscosity changes referentially to reservoir pressure values below the bubble point.

From Figure 4.12, we can see that several EOR chemical field projects (dark and light dots) surround the Field A cases that were evaluated: the micellar polymer floods in Lawrence field in Illinois (Gogarty, 1976; Thomas, 1974) and North Burbank in Oklahoma, two potential close analog (Bradford et al., 1980; Lorenz, 1986; Trantham, 1983). Table 4.6 shows the basic reservoir parameters of the Lawrence (also known as Lawrence-Robinson) field and North Burbank extracted from the database.

The main reason to identify the Lawrence and North Burbank fields as possible analogs is the water salinities that exist in both fields. Lawrence field and North Burbank report water salinities of 33,000 to 50,000 ppm and 75,000 to 115,000 ppm, respectively. However, from the point of view of reservoir properties, North Burbank seems to be quite similar to Field A (see Tables 4.5 and 4.6).

Once this kind of match is identified, the next step dedicates a reasonable effort to searching for and gathering as much documentation as possible by describing the details of the project in order to learn from past experiences, reducing potential uncertainties associated with a particular EOR application at very early stages of evaluation. Be aware that this is in no way meant to guarantee applicability; instead it provides referential information as to how experience can be used for a project of interest.

TABLE 4.6 Basic Reservoir Parameters of Possible Closer Analog (High-Salinity Reservoirs) for Field A

Possible Analog	Porosity (%)	Temperature (°F)	Pressure (psi)	Permeability (md)	Oil Gravity (°API)	Oil Viscosity (cp)
Lawrence Field, IL	20	65	300	180	34	24
North Burbank, OK	16	120	1,350	52	39	3

In this particular case study, the evaluation went through more intensive data gathering for the North Burbank field, given the similarities found in different reservoir properties and conditions compared to Field A (e.g., formation and injection water salinities, average permeabilities, decreasing permeabilities with depth, rock mineralogy, and the Dykstra-Parsons coefficient). It is important to mention that North Burbank reported a successful micellar polymer flood and a polymer flood at a commercial scale that were well documented (Bradford et al., 1980; Clampitt and Reid, 1975; Kleinschmidt and Lorenz, 1976; Lorenz, 1986; Trantham, 1983; Trantham et al., 1978).

The information published has proved very valuable for developing conceptual numerical simulation studies to evaluate the technical and economical feasibility of surfactant-polymer flooding in Field A. In other words, we borrowed the North Burbank chemical formulation to carry out preliminary studies by performing multiple sensitivities on key variables such as surfactant–polymer adsorption, interfacial tension reduction, and surfactant–polymer concentration.

From this simulation we identified and ranked critical variables that contributed to the methods used for ongoing laboratory efforts (e.g., polymer viscosity and its relationship to the injectivity and critical surfactant adsorption that may impact final recovery factors). Therefore, the authors believe that this simulation approach definitively contributes to the management decision process of EOR evaluation prior to expensive and time-consuming studies (e.g., laboratory studies, field data gathering, and/or reservoir studies). For more details on the different numerical simulation approaches, please refer to Chapter 3, Simulations and Simulation Options.

4.4.2 Analyzing a Portfolio of Oil Reservoirs

The authors have used this advanced screening methodology in multiple portfolios. Indeed, the methodology was developed in response to corporate exploration and production planning needs to evaluate hundreds of crude oil reservoirs in Venezuela in order to identify EOR potential in different regions of the country (Alvarado et al., 2002; Hernández, Liscano, et al., 2002). Manrique and colleagues (2003 and 2006) documented two cases of portfolio analysis in crude oil reservoirs in Tia Juana (Venezuela) and Wyoming (United States), evaluating the potential of CO_2-EOR in multiple reservoirs. The following describes one of the widely documented approaches used to estimate the EOR potential in Venezuela.

This analysis of a particular portfolio of reservoirs in South America is based on the same criteria explained in Section 4.4. Specifically, a separate database from an operator (major or independent) is reorganized to generate the required input file to automatically run the screening and quickly identify reservoir typologies. Most of the common recovery processes are identified to estimate the value of a particular technology, and potential analog are used to justify additional studies.

Figure 4.13 is an example of a portfolio of carbonate reservoirs using the expert map shown in Figure 4.10. Additionally, representative carbonate reservoirs of the Middle East (Abu Safah—Arab D, Ghawar, Haft Kel, Ikiztepe, Issaran, Qarn Alam, Sabiriya, and Shaybah fields) were included in the analysis to evaluate if these fields showed some similarities to the portfolio of reservoirs under evaluation.

Based on the location of the fields included in the analysis, the results suggest that Middle East reservoirs are shallower and less permeable (e.g., less fractured) than the fields evaluated. This analysis may also include different degrees of granularity for a more detailed analysis than that described in the previous section. Additionally, the portfolio of fractured carbonates reservoirs analyzed shows the potential for CO_2 injection (either continuous or in a WAG mode), assuming that both natural and anthropogenic CO_2 sources will be available based on a separate independent study (perhaps by looking at the associated soft issues).

Another example of a portfolio of reservoir analyses is the evaluation of different Venezuelan basins to support the integrated field laboratory (IFL) philosophy implemented by the Venezuelan Oil Industry (Petróleos de Venezuela, or PDVSA) from 1998 to 2002 (Alvarado et al., 2002; Hernández, Chacon, et al., 2002a, 2002b, 2003; Manrique et al., 2000a, 2000b). Specifically, the IFL's main strategy was to evaluate EOR technologies in different reservoir typologies in highly integrated pilot projects. The VLE area was selected for the evaluation of immiscible WAG processes with hydrocarbon gases.

The VLE area is representative of a large number of deep (>10,000 ft) light-oil reservoirs in the Maracaibo Lake basin with similar reservoir characteristics ("reservoir typology"). Additionally, Venezuela has over 14,000 MMSTB of oil currently in place in reservoirs with similar conditions. The selection of the VLE-305 field (offshore Maracaibo Lake) was based on EOR screening studies for WAG flooding that were obtained from an analysis of successful WAG projects worldwide; analytical, experimental, and numerical simulation studies; and the availability of water, gas, and surface facilities. The WAG pilot project at VLE-305 represents a good example of an EOR evaluation from screening to field implementation and monitoring that is well documented in the literature (Alvarado et al., 2002; Alvarez et al., 2001; Hernández, Chacon, et al., 2002a; Manrique et al., 1998, 2000b; Stirpe et al., 2004).

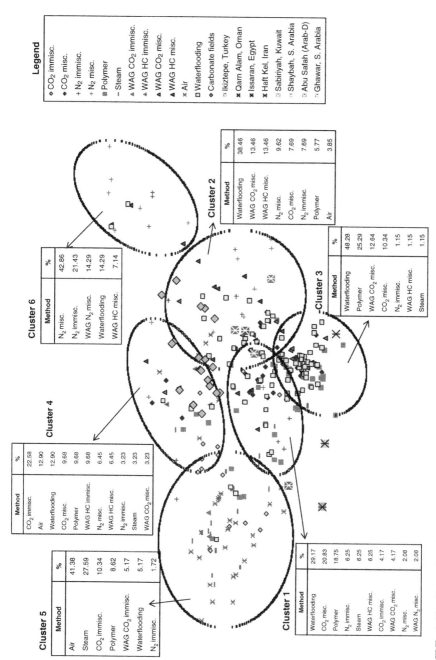

FIGURE 4.13 Example of screening of a portfolio of carbonate reservoirs (*diamonds*) compared to representative carbonate fields in the Middle East.

4.4.3 Other Applications

Additional examples of the advanced screening methodology that is described in this chapter focus on the sensitivity analysis of dynamic parameters such as reservoir pressure and crude oil viscosity (e.g., when the reservoir pressure goes below the bubble point, or P_b). Figure 4.14 shows an example of a pressure decline in a reservoir, B, under evaluation. The following reservoir properties were used for the analysis:

- Porosity = 22 percent
- Permeability = 500 md
- Reservoir temperature = 170°F
- Crude oil gravity = 32° API
- Reservoir pressure evaluated: 3,000 psi to 500 psi (P_b = 1,000 psi)
- Crude oil viscosities = 10 cp above P_b and 24 cp below P_b

Reservoir B falls into cluster 4 of the expert map shown in Figure 4.14. As pressure declines, it approaches possible analogs under different recovery strategies (e.g., waterflooding and polymer flooding). The pressure decline from 3,000 psi to 1,000 psi follows a linear trend until the reservoir pressure hits the bubble point, P_b = 500 psi. At this point the trend changes as a result of a higher viscosity of the crude oil at reservoir conditions. It is important to mention that reservoir B at low pressure (500 psi) and higher crude oil viscosity (24 cp) at reservoir conditions yields potential analog under immiscible CO_2 injection (light diamonds in Figure 4.14).

We consider this example very useful for explaining the impact of reservoir pressure on IOR/EOR decision-making processes. Assuming that reservoir B is produced under natural depletion and that a water injection project starts below the bubble point pressure, it is highly probable that poor displacement efficiencies will develop due to the presence of free gas in the formation and thus lower final recoveries.

Finally, pressure decline using the proposed advanced screening method has also been used to evaluate potential reservoir development plans; this was done during exploratory appraisal that was based on limited data. It provided preliminary options for the development a particular field.

4.4.4 CO_2 Expert Maps

Given the importance of and the increased interest in CO_2-EOR from both natural and anthropogenic sources, the advanced screening was adapted to evaluate CO_2-EOR projects (Manrique et al., 2003). The CO_2 expert map was developed using the same approach described earlier in the chapter. Once database collation and quality control are carried

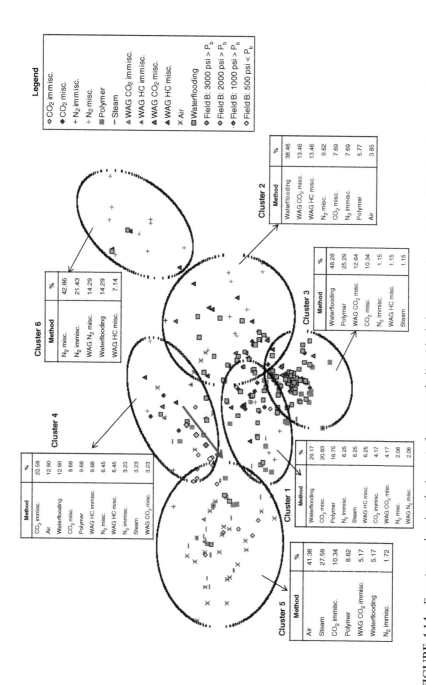

Legend

◇ CO_2 immisc.
◆ CO_2 misc.
+ N_2 immisc.
+ N_2 misc.
■ Polymer
— Steam
▲ WAG CO_2 immisc.
▲ WAG HC immisc.
▲ WAG CO_2 misc.
▲ WAG HC misc.
✕ Air
■ Waterflooding
◇ Field B: 3000 psi > P_b
◇ Field B: 2000 psi > P_b
◇ Field B: 1000 psi > P_b
◇ Field B: 500 psi < P_b

Cluster 6

Method	%
N_2 misc.	42.86
N_2 immisc.	21.43
WAG N_2 misc.	14.29
Waterflooding	14.29
WAG HC misc.	7.14

Cluster 2

Method	%
Waterflooding	38.46
WAG CO_2 misc.	13.46
WAG HC misc.	13.46
N_2 misc.	9.62
CO_2 misc.	7.69
N_2 immisc.	7.69
Polymer	5.77
Air	3.85

Cluster 4

Method	%
CO_2 immisc.	22.58
Air	12.90
Waterflooding	12.90
CO_2 misc.	9.68
Polymer	9.68
WAG HC immisc.	9.68
N_2 misc.	6.45
WAG HC misc.	6.45
N_2 immisc.	3.23
Steam	3.23
WAG CO_2 misc.	3.23

Cluster 3

Method	%
Waterflooding	48.28
Polymer	25.29
WAG CO_2 misc.	12.64
CO_2 misc.	10.34
N_2 immisc.	1.15
WAG HC misc.	1.15
Steam	1.15

Cluster 1

Method	%
Waterflooding	29.17
CO_2 misc.	20.83
Polymer	18.75
N_2 immisc.	6.25
Steam	6.25
WAG HC misc.	6.25
CO_2 immisc.	4.17
WAG CO_2 misc.	4.17
N_2 misc.	2.08
WAG N_2 misc.	2.08

Cluster 5

Method	%
Air	41.38
Steam	27.59
CO_2 immisc.	10.34
Polymer	8.62
WAG CO_2 immisc.	5.17
Waterflooding	5.17
N_2 immisc.	1.72

FIGURE 4.14 Expert map showing the impact of pressure decline (shaded arrow in cluster 4) for a field (B) under evaluation.

out, the next step is to process the data to generate a knowledge map based on 80 CO_2 floods. Although more than 15 variables were initially considered, in order to have a large number of records available for the analysis, the same 6 variables were selected to generate the maps. The selection was based not only on the importance of those variables to form well-defined clusters but also on the reduction of redundant information, based on the correlation analysis.

Given the fairly good correlation between temperature (T) and the true vertical depth (TVD), it was decided to exclude the reservoir temperature in generating the knowledge map. Additionally, and after much thought, it was found that crude oil viscosity did not improve the generation of the CO_2 expert map. Therefore, the five variables selected to generate the knowledge maps of international CO_2 floods were crude oil gravity (API), reservoir depth (TVD), average porosity (ϕ), average permeability (k), and reservoir pressure (P) at the beginning of the CO_2 project. The only difference in the CO_2 expert map is that the clustering of reservoir typologies was done based on formation type or reservoir lithology.

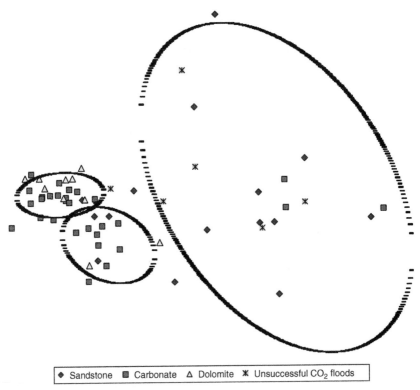

♦ Sandstone   ■ Carbonate   △ Dolomite   ✱ Unsuccessful CO_2 floods

FIGURE 4.15   Advanced screening (expert) map for CO_2-EOR by lithology.

After applying a clustering algorithm to the projection, three clearly defined clusters were determined (Figure 4.15). These clusters represent three mixed reservoir typologies—that is, each cluster is made up of different reservoirs where various type of lithologies and CO_2 floods have been implemented (continuous injection or WAG mode). Reservoir lithologies were classified in three main groups: sandstones (diamonds), carbonates (squares), and dolomites (triangles). Figure 4.15 also shows six unsuccessful CO_2 floods for different reasons, as documented in the literature.

Figure 4.16 illustrates CO_2-EOR screening for a portfolio of sandstone and carbonate reservoirs (Fields E) in the United States, identifying potential analog in Louisiana (Bayou Sale and Quarantine Bay), Texas (Penwell), and a failed CO_2 flood in Mississippi (Heidelberg). The figure also shows (in the screened box at the top right side) three of the five polar axes representing the magnitude of variables such as permeability, crude oil gravity (°API), and reservoir depth.

For example, field case E1 represents a carbonate reservoir (circle) where sensitivity on reservoir permeability was considered. The high permeability value ($k = 1,000$ md) falls in the cluster dominated by sandstone reservoirs, while the low permeability value ($k = 0.1$ md) is displaced to the left and close to the cluster dominated by CO_2 floods implemented in carbonate and dolomitic formations. Finally, this methodology has been used for property evaluation and acquisition as well as for the evaluation of potential CO_2-EOR storage in the United States and abroad (Manrique and Wright, 2006; Velásquez et al., 2006).

Once screening (conventional, geologic, and advanced) is completed, the next step is to predict the oil production of selected EOR processes via analytical or numerical simulation. Frequently, project evaluation does not have enough laboratory and field data to run simulations. In such cases, input data are generated from field data (e.g., pore volume injected, WAG ratios, chemical formulation, and adsorption data); correlations, closer analog, and/or reservoirs from the same geologic formations (geologic screening) can be identified in the EOR screening phase.

Finally, the major utility of advanced screening is the ability to rapidly screen projects and rank enhanced oil recovery options in a particular field without expert bias or by previous field experiences.

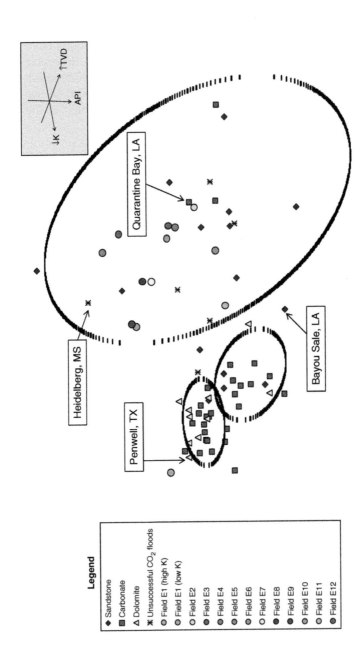

FIGURE 4.16 Example of CO_2-EOR screening of a portfolio of reservoirs (Field E) that identifies potential analog. Field Case E1 is a carbonate reservoir (high k = 1,000 md, average value including fractures, and low k = 0.1 md representing matrix permeability).

Evaluating Soft Issues

5.1 INTRODUCTION

It is a fact that only a few enhanced oil recovery (EOR) screening studies in the literature thoroughly evaluate soft or nontechnical issues as an important component of EOR screening strategies or exercises (Alvarado et al., 2008; Bu et al., 1993; Manrique et al., 2008). Recent examples are worth some attention in connection with CO_2-EOR storage studies, as analysts attempt to incorporate both technical and nontechnical issues (e.g., health, safety, legal, and the public's perception) into the overall planning of carbon-capture storage projects (Christensen, 2006; Flanery and McCarty, 2008; Garland, 2007; Ghomian et al., 2006; Hamilton, 2009; Imbus et al., 2006; Manrique and Araya, 2008; Xie and Economides, 2009). This naturally arises in greenhouse storage analyses as a result of the social and environmental sensitivities associated with the issue at hand and the constant media exposure.

You should be aware that the value of evaluating soft, or nontechnical, issues is that it prevents having to carry out detailed analyses on nonfeasible EOR processes—for instance, because of the lack of injection fluids or offshore restrictions—before data must be made available for much more elaborate analyses. For unproven technologies, decisions may involve developing embryonic technologies. These cases are not amenable to the same type of "hard" analysis that uses economic indicators as objective functions or values.

It is not uncommon to see evaluation results where the technical feasibility of a particular EOR process is (or has been) demonstrated, only to encounter later the conundrum of identifying potential sources of injectants (e.g., CO_2, water, or gas for steam injection) that are quite unlikely. This is also true for proven technologies that may pose barriers to fully developing oil resources due to environmental constraints, regulatory frameworks, or lack of natural gas resources to generate steam, just to mention a few common examples.

This chapter provides some examples to give you a general idea of the meaning of soft and nontechnical issues that can impact or delay, or have impacted or delayed, the implementation of EOR projects, as well as property evaluations and acquisitions. The examples we use are centered on some of the main themes regarding soft issues that make a significant difference in the decision analysis of EOR projects.

5.2 INJECTION FLUIDS

One of the most common soft issues that arise in property evaluation and acquisitions, as well as in the implementation or expansion of steam injection projects in heavy oil reservoirs, is the lack of natural gas and/or water to generate steam. Chemical EOR methods are also susceptible to injection fluid availability, specifically water quality for injection. Here are some examples:

Water availability for steam generation. In this case, a preliminary evaluation of potential sources and volumes to justify a sustainable steam injection project is recommended (Dell et al., 2008; Horne et al., 2002; Veil et al., 2009). If water will be produced from aquifers, it is important to estimate water reservoir depth and productivity. General information can be obtained through environmental agencies, geologic surveys, commercial databases (e.g., well logs), and surveys of open literature.

Water quality for steam generation. The two main considerations in these cases are high-quality water cost and disposal of produced water. In some cases, water treatment can be too costly and thus negatively impact the economy of steam injection projects. Additionally, water treatment can generate important volumes of waste (harsh) waters that need to be disposed of, adding operational costs (OPEX) to projects (Beaudette-Hodsman et al., 2008; Dell et al., 2008; Garbutt, 1997; Neff and Hagemann, 2007; Pedenaud and Dang, 2008; Webb et al., 2009). Disposal well permitting, well costs, and identifying deeper formations where the water can be disposed of also must be considered, at least preliminarily.

Water quality for chemical EOR methods. Although water used for EOR does not need to be fresh, special attention must be paid to the costs of water treatment, and to its operational implications in the early stages of any evaluation (Chang et al., 2006; Hernández et al., 2001; Meyer, 1988; Royce et al., 1984; Wang et al., 2009; Zainal et al., 2008).

Natural gas availability. This has become one of the key issues for steam flooding projects. Limited gas volumes combined with high market demands and volatile energy prices have contributed to the decline of steam injection projects in the United States during the last two decades (Aalund, 1988; Moritis, 1998; Moritis, 2008; Noran, 1978). Therefore, how many natural gas resources are available (e.g., associated gas, gas pipelines, coal bed methane, or CBM, production near the field under study) is a prime factor in whether the conditions are right for a steam injection project, even when there are positive screening results (Vassilellis, 2009). Even in cases where natural gas resources are available, higher costs due to market competition can render steam injection projects economically unattractive (Freedenthal and Taylor, 1989).

The availability of natural gas to develop the Canadian oil sands has been addressed in recent years. Canada, however, may not have enough natural gas to generate the required steam volumes necessary to develop its vast oil sand resources using steam-based processes (e.g., cyclic steam injection and SAGD). To mitigate the lack of natural gas, several options have been proposed. Bitumen gasification, combined with CO_2 storage in deep saline aquifers, and nuclear energy are two alternatives for replacing natural gas for steam generation (ACR, 2004; Asgarpour, 2009; Bersak and Kadak, 2007; Garnier and Cupcic, 2002; Geddes and Curlett, 2005).

5.3 CARBON DIOXIDE

Carbon dioxide (CO_2) is a classic example of constrained gas injection availability. CO_2 storage has gained increasing interest in the discussion of CO_2-EOR projects. Several screening and feasibility studies of CO_2-EOR consider CO_2 from natural or anthropogenic sources. Abundant literature on international CO_2 studies is available, but a limited number of references are provided in the literature (Algharaib, 2009; Gaspar et al., 2005; Hamilton, 2009; Mohan et al., 2008; Scharf and Clemens, 2006; Wo et al., 2008; Wo et al., 2009). However, the reality is that in the vast majority of cases, CO_2 is not available, and even if it is the sources are far from the EOR reservoir candidate.

This situation increases the costs associated with compression and transportation. If CO_2 is to be captured from industrial sources, the costs of actual technologies (e.g., capture, compression, and transportation) can be prohibitive and can therefore make CO_2-EOR infeasible in the near future, especially in mature fields. CO_2 economics will be briefly discussed in Chapter 6, Economic Considerations and Framing.

Regarding the revitalization of mature fields by CO_2-EOR, if CO_2 sources are not available at reasonable costs and in a reasonable period of time, there is a high probability that an important number of reservoirs will be abandoned, losing the window of opportunity ("timing") to increment oil recoveries. In other cases, CO_2 may become available for depleted reservoirs too late to justify its injection because the decreased pressure would allow immiscible CO_2 floods only. Immiscible EOR processes are characterized by low ultimate recovery factors—generally below 5 percent (Christensen et al., 2001)—and, in particular, CO_2 injection. Given the capital expenditures (CAPEX) associated with this kind of project, they make the whole idea quite unattractive in the face of limited opportunities.

It is paramount to recognize that, as of today, CO_2-EOR is viable through the existence of vast CO_2 natural resources in the United States (Ghomian et al., 2008; Holtz, 2008; Hustad, 2009; Imbus et al., 2006; Manrique and Araya, 2008; Manrique et al., 2007). Indeed, Hustad (2009) estimates that approximately 70 percent of global oil production from CO_2-EOR projects is coming from the Permian Basin in Texas, where vast natural CO_2 reservoirs are available.

Another important fact or in CO_2-EOR in the Permian Basin (New Mexico and Texas) is the CO_2 pipeline developed to support these projects. The CO_2 pipeline was constructed during the 30-year period (Hustad, 2009) in which the CO_2-EOR projects came online. Therefore, when performing EOR screening studies to determine technical feasibility of CO_2-EOR, it is important to be realistic before investing time and project resources. It is important to identify the sources of CO_2 (natural vs. anthropogenic) and their distance from the reservoir under consideration as early as possible during screening evaluations, and to exercise good judgment.

5.4 GEOGRAPHIC LOCATION

As we just said, CO_2 availability from natural sources in the United States represents a good example of how geographical location can play a major role in decision management for the selection of a particular recovery process that involves EOR methods. The following are some examples of successful EOR decision management strategies as a result of a particular geographic location: Alaska, the North Sea, Montana, North Dakota, and South Dakota.

Alaska. The north slope of Alaska did not have access to gas markets when large reservoirs were put into production. Therefore, recovery strategies have been based on pressure maintenance, and EOR technologies have been supported fundamentally by gas and water injection. In this example, decision management strategies focused on reservoir development plans that maximized and optimized the resources available.

EOR technologies, such as miscible gas injection (miscible injectant, MI), water-alternating-gas (WAG), low-salinity water injection (LoSal), and in-depth diversion systems (BrightWater) are only a few of the many recovery strategies implemented on Alaska's north slope (Ding et al., 2009; Ohms et al., 2009; Panda et al., 2009; Patil et al., 2008; Platt, 2008; Seccombe et al., 2008; Szabo, and Meyers, 1993).

North Sea. The main drainage strategy in the North Sea and in most, if not all, offshore fields has been pressure maintenance by gas and water injection. It is well known that EOR strategies are limited in offshore operations. Decision management strategies based on pressure ("energy") maintenance by injection of gas, water, and/or WAG close to or well above bubble point pressures have allowed North Sea operators to extend the productive life of fields by implementing reservoir depressurization (blow-down strategies).

This change in the field production mode from oil to gas, with significant volumes of produced water, not only will prolong the lifetime of offshore installations, delaying their decommissioning, but will also lead to increased oil and gas recoveries, opening new business opportunities in the region (Awan et al., 2008; Beecroft et al., 1999; Blanksby et al., 2005; Boge et al., 2005; Chekani and Mackay, 2006; Drummond et al., 2001; Gallagher et al., 1999; Quint, 1999).

Other Areas. The recovery locations in Montana, North Dakota, and South Dakota are in remote areas where gas for injection is not commonly available for large projects. Specifically, low-permeability ($K \cong 10$ md) and deep ($> 8,300$ feet) dolomite light-oil ($31–33°$API) reservoirs of the Red River formation in these three states are candidates for gas flooding on the basis of conventional and advanced screening criteria.

During the 1980s, however, evaluation of high-pressure air injection (HPAI) started because of the lack of other injectants such as nitrogen (N_2) and/or CO_2 (natural or from industrial sources) or their costs. HPAI projects have been steadily increasing since then (Moritis, 2008).

Since HPAI was pilot-tested successfully in Buffalo Field (South Dakota), the risks and uncertainties involved with subsequent implementations in similar fields have been drastically reduced. The increased confidence of local operators has made a positive impact on thermal EOR in carbonate formations in Montana, North Dakota, and South Dakota (Gutiérrez et al., 2008a, 2008b; Kumar et al., 2008; Turta and Singhal, 2001).

It is important to note that, during 2000, CO_2 became available as part of the CO_2 capture implemented by the Dakota Gasification Company at their Great Plains synfuels plant in North Dakota (Dakota Gasification Company, 2009). However, the CO_2 is currently being transported by pipeline to Weyburn oil field in Saskatchewan (Asghari et al., 2007; Malik and Islam, 2000), where it is fully utilized for CO_2-EOR. Therefore, and despite the availability of CO_2, HPAI projects have dominated EOR activities in the region during the last decade. The distance from the CO_2 source to the CO_2-EOR reservoir candidates and the high CAPEX associated with CO_2 floods may not compete with HPAI projects and the experience gained in the last two decades.

To summarize, when performing EOR screening evaluations, it is important to consider the geographic location and the resources (e.g., injectants) available for the asset/reservoir under evaluation. This will help you identify EOR opportunities that can be realistically put forth, technically and economically, in a reasonable period of time. It will also contribute to corporate portfolio analysis and decision risk management strategies. Waiting for sources of "nice to have" injectants (e.g., CO_2) may reduce EOR potential and thus the final recovery factors for a particular oil reservoir.

5.5 OTHER SOFT ISSUES

The soft, or nontechnical, issues just discussed are a few of those that the authors have consistently encountered during consultancy and engineering studies. It is important to note that soft issues can be different for each EOR technology and relative to its applicability in a particular scenario. Although the list of possible situations can be extremely long, here are a few additional examples of soft issues when performing EOR screening and/or evaluation studies.

5.5.1 Regulations

Environmental regulations have to be considered in every EOR evaluation. However, these regulations are specific to a region. They also differ depending on whether the application is for an offshore or an onshore EOR (Garland, 2005; Godec, 2009; Godec et al., 1993; Holliday,

2009; Marinello et al., 2001; McMillen, 2004; Minami et al., 2003; Shyam et al., 1995). Therefore, proper environmental assessment studies are critical to support the implementation of EOR technologies, considering the potential impact on the ecosystem, economics, and social aspects. Public perception (e.g., safety) and application for environmental permits (e.g., disposal wells), if required, are important factors that may delay EOR field applications.

5.5.2 Public Opinion

Public acceptance, legal aspects, and understanding CO_2 storage are among the main tasks ("soft issues") for all key players (e.g., mining and utility sectors, oil and gas industry, regulatory agencies) involved in CO_2-EOR storage projects. The focus must be on demonstrating that CO_2 can be safely stored in geologic formations over very long periods of time (Flanery and McCarty, 2008; Garland, 2007; Imbus et al., 2006; Manrique and Araya, 2008).

5.5.3 Expiration Dates

Lease or concession expiration dates may play a role in EOR decision risk management strategies. Some operators have to factor in the timing of implementation and project economics (e.g., the rate of return and revenues). Although in some countries lease extensions can be obtained without major problems, often it is not possible to delay or to cancel EOR project implementation because of their high capital exposure and associated risks.

Several authors have addressed this topic, not specifically in EOR projects but in oil and gas projects in general, including the environmental implications of lease transfers during nationalizations (Armstrong et al., 2008; Dezen, and Morooka, 2001; Laughton, 2003; Lazo et al., 2007). Generally top management at the corporate level is well aware of lease terms and agreements, and this information should be shared at different levels because of the importance of timing for implementing EOR projects.

5.5.4 New Technologies

Implementing new enhanced oil recovery technologies that come from research and development (R&D) generally must overcome physical and/or operational limitations (e.g., logistics) that make projects difficult to implement or that become too costly. Taking technologies from the laboratory for scaleup to pilot projects involves a high degree of risk and potentially high capital costs (e.g., new wells, monitoring). However, associated risks need to be measured against the upside potential (promise) of the technology. We believe that the best experimental result is sometimes the field itself if all of the technical and economical considerations have been properly addressed.

5.5.5 Public Information

The perception created as a result of various published information—peer-reviewed articles, conference proceedings, and so on—can generate biases that lead to suppressing or incorrectly encouraging unbalanced views of EOR potential. Advertising EOR technologies does not always guarantee success. However, little information about a particular EOR application does not necessarily mean that an EOR method must be ruled out if the screening turns out to be positive. Chemical EOR is paradigmatic in this sense.

A few recent field experiences have been documented in the literature. For example, the last *Oil & Gas Journal* EOR survey reported by Moritis (2008) shows only two planned projects (in Texas) and two ongoing projects (in Oklahoma). Misinterpretation of this well-respected survey may create the wrong perception of the actual state of affairs regarding chemical flooding, as reflected by the number of ongoing lab and pilot project designs as well as field-scale projects. In reality, there is more ongoing activity than that published in the open literature.

Although the authors believe in the importance of technology transfer and learning from previous field experiences, there are several good reasons for certain operators to keep information and field results out of the public domain. Here are some of them:

- Reduced pressure on chemical product demand and thus surfactant prices and supplies, especially during periods of high oil prices.
- Having a successful chemical flood not only adds value to a particular asset but also provides the opportunity to develop a plan for property acquisitions following the U.S. Securities and Exchange Commission definition for "reservoir analogs" (SEC, 2008). In other words, EOR potential validated in a particular reservoir brings the opportunity to acquire neighboring or nearby fields producing from the same formation, with consequent optimization of capital investment for companies assuming the risk of testing EOR technologies, despite the volatility of energy prices.

5.5.6 Cost of Imports

When considering the evaluation of chemical EOR methods, such as surfactant-polymer (SP) and alkali-surfactant-polymer (ASP), it is important to understand the impact that import costs of chemical additives have on pilot or field-scale projects. These costs can be as much as 30% of the project's total expense. Therefore, the evaluation of surfactant and/or polymer (for EOR applications) manufacturing capabilities within or near the country or region where the project is being considered will definitively contribute to business decision strategies. Although imports of chemical additives are common for projects of this type, and should not be considered a limiting factor for EOR implementation, cost reduction associated with local manufacturing capabilities is always welcome and can reduce the expense of EOR implementation as well. Finally, import costs (e.g., customs, taxes, permits, nationalization) are country specific.

5.5.7 Labor

One important benefit that is generally overlooked in EOR projects is direct and indirect employment opportunities that a particular project may bring to an area. Direct and/or indirect jobs created will contribute to local communities and to the economic wealth of a country and/or region. This can play an important role in decision making at both corporate and government levels.

5.5.8 Summary

In closing, it is important to highlight that each EOR technology may have its own set of soft, or nontechnical, issues relative to its applicability under particular circumstances. Soft issues may include the following:

- Health, safety, and environmental (HSE) issues
- Regulatory framework or controls
- Legal and liability issues (e.g., hydrocarbon legislation)
- Government conditions (e.g., terms for leases and concessions)
- Public perception and a combination of the preceding issues

Continuous changes in both the environmental and the legal regulatory framework recently observed in various countries may negatively impact the evaluation and commercialization of EOR technologies, leading a loss of new business and job opportunities. Therefore, it is important to consider and investigate potential nontechnical issues that may affect decision risk management of particular EOR technologies in the early stages of evaluation, before time-consuming and expensive studies are started.

Economic Considerations and Framing

6.1 INTRODUCTION

This chapter covers two intertwined aspects of the same problem: framing and economics. Framing has already been mentioned several times in this book, but it is formalized in this chapter. We cover basic ideas on economic/financial evaluations and draw a bigger picture in the context of the real business setting. The material is not intended to be comprehensive. You should make sure that you will be using appropriate financial models for your evaluations. At the same time, make sure that you fully comprehend your organization's decision-making style.

6.2 FRAMING DECISIONS

We start by discussing a number of useful tools for framing that serve as decision enablers. One tool we often use is the influence diagram to

Enhanced Oil Recovery
DOI: 10.1016/B978-1-85617-855-6.00012-7

graphically represent the connection between critical decision-making variables and the corresponding EOR process. The purpose of the graphical decision model representation is twofold. First, this order of presentation will allow us to elaborate on influence diagrams. Second, we will provide a graphical anchor to help you keep in mind the guiding design principles for EOR processes.

It is important to understand that analysts do not generally make decisions; instead, decision makers, whether they be managers, investors, or any empowered individual or organization, make the decisions because they can in fact commit the resources. This point cannot be overstated because a frequent reason for failed decisions arises from a misunderstanding as to who should be addressed in the decision analysis exercise. Analysts recommend and advise, while decision makers commit financial resources and therefore must be empowered to do so.

Rational decision making is only possible when the two functions keep a degree of separation and often independence. It is with the help of decision makers that analysts build a framework to assist the decision-making process. Soft issues play a role in decision making, but quantifiable objective functions have to be selected to enable rational decision making. This complex dynamic is part of a typical decision and risk management effort. There is more on this to come later.

6.2.1 What Is a Decision?

The first question we need to answer is, What is a decision? We will follow Skinner (1999): "A decision is a conscious, irrevocable allocation of resources with the purpose of achieving a desired objective." We equate "conscious" to "rational," meaning that you are thinking about the process of making a decision, most likely with a clear, well-defined objective in mind; in other words, you decide through a rational process. Your eloquence, experience, and, consequently, your biases will not make a decision rational.

The main function of the support team, consultant, or analyst is to provide alternatives for the basis of the framed decision problem so the best decision can be made—again by the decision maker. You must make one distinction, however, to cope with negative outcomes. If you have properly built a decision-making process and have followed a rational procedure, the outcome does not impact the validity of the decision. The best way we can explain this is with an example.

Most people understand that the odds of buying a winning lottery ticket are quite low. In this sense, anyone who follows a well-designed decision-making procedure would tell you that buying a lottery ticket is not a guarantee of winning the lottery. Now the question is, If you buy the ticket and actually do win the lottery, was buying the ticket a good decision? It turns out that rational decision making would indicate that even if your winning the lottery does not make up for the bad decision to buy the ticket.

Let us now expand on something we said in the introduction to this chapter. Say an opportunity presents itself as a result of an analysis that was conducted by an organization's staff—whether a junior engineer or an integrated asset team—or a review of reserves (the case at hand is almost irrelevant). Or maybe the idea came about from reading conference proceedings or attending a seminar on a new technology (which has been the authors' experience more than once). Let us now say that the idea is further elaborated on to become a concept, and either a back-of-the-envelope calculation or a quick lab test (or even a simplified numerical simulation) supports the potential of this possible opportunity.

Although it might sound as if this is not a good idea, on the contrary, we encourage this process, but we do caution you on such a course of action. It turns out that you, and possibly your colleagues, are convinced of the potential of this strategy. You then proceed to contact a supervisor, manager, or business partner. You build the case using your good reputation, eloquence, and solid technical background and manage to spread your contagious enthusiasm because you cannot believe you will meet with any resistance to your idea. Your interlocutor agrees with your proposal and devises a plan of action.

At this point, you believe you have managed to convince the decision maker and are certain that the project will come to fruition. By the way, it might be a business partner or even a supervisor. This latter scenario is actually very risky because of the biases introduced prior to analysis that can drive a self-fulfilling prophecy, meaning that a person empowered to commit resources actually forces the team to "prove" his or her point. (This also happened to these authors at some point, but back to the story.)

What is wrong with this picture? If you obtained resources to further elaborate on your idea and even to carry out a project financial analysis, you are well under way to creating your project. It all sounds good, and resources have been committed. But wait! The resources we are talking about are for a decision within a framed project—and are you sure you talked to the decision maker? Can this person or organization decide on the volume of resources that will lead to deploying a process or even committing to field surveys or detailed laboratory experiments? If the answer is no, then you have been most likely talking to a member of your team or a well-intentioned supervisor, partner, or anybody else—but not to the decision maker.

6.2.2 Decision Objective Function and Financial Indicators

The point of the preceding discussion is that rational decision making is critical; framing the process correctly will make all the difference.

Working on the right problem is perhaps the most important aspect of decision making, particularly in EOR. The decision maker defines the frame by providing preferences and decision criteria. The former probably reflect the degree of risk aversion and policies, while the latter will yield the objective or value function. This value function should be a quantifiable function that the team can ascertain. Typically, a financial indicator or indicators are used as objective functions, but these are not the only ones that can be used.

A time-value-of-money function, such as *net present value* (NPV) and *internal rate of return* (IRR), is often the objective function because it allows you to compare projects. If you are unfamiliar with these concepts, we recommend that you pick up a reference on financial project evaluation so these concepts become ingrained in your memory; they are *that* important.

The concept of the time value of money essentially reflects the fact that the cash flow today, or the present value, is not the same as that same amount of money in the future, or the future value. The reason for this is that idled resources lose value if they go unused because opportunities to invest and make a profit are always available, even if all you do is put your money in a savings account in a bank. To compare cash flows over a period of time, we resort to the concept of present value by calculating the value of successive cash flows over a number of years using the concept of IRR. Let us first consider rate of return. If we put money in a bank account, a return will be produced over the investment (deposit) every year.

Although this example is an extremely simple one, it clarifies the whole concept. IRR is the value of the discount interest rate at which the cash outflows (or derogation) will equal the inflows at a particular point in time. The rule of thumb is that the higher this rate is the better. A number of projects compared on equal bases (including time) can in principle be sorted out from better to worse by comparing IRR. However, IRR does not allow one to identify the net return on investment in monetary terms. Analysts often base their analyses on NPV because volumes of resources are actually compared. If you have the successive net cash flows (NCF) $X_0, X_1, X_2, \ldots, X_n$, then NPV is calculated as

$$NPV = \sum_{t=1}^{n} \frac{NCF_t}{(1+k)^t}$$

where t is the year for each NCF, $X_0, X_1, X_2, \ldots, X_n$, while k is the annual discount rate. The net cash flow is calculated every year as

$NCF(\text{in } \$) = \text{Revenue} - \text{Capital investment} - \text{Operating expenses}$

Revenues arise from sales, just as hydrocarbon reservoirs come from the sales of oil and gas. Two other important concepts are *payout time,* or the time needed to recover the investment, and *investment efficiency*, which is the ratio of the net cash flow to the total discounted investment (Satter et al., 2008).

If you have never dealt with these concepts before, think of them as a way to compare sums of money in the cash flow operations at different times. As we explained before, the fact that there is a rate of return, which is a simple fact in the business world (hence the expression "Time is money"), makes it impossible to compare money flow in different years. There is a direct consequence of the time value of money for IOR/EOR. It turns out that the faster you produce the "same" resources, everything else being equal, the higher the NPV. By the way, this is a good incentive to "accelerate" production, meaning that the faster you produce the oil, the higher your potential NPV will be.

This can be very risky for EOR decisions because by using only one financial indicator, you might overemphasize IOR operations that will not necessarily lead to the maximization of final recoveries. Although it satisfies short-term positive cash flow needs, production acceleration by means of formation stimulation (e.g., acidizing, hydraulic fractures, perforation campaigns, sand coproduction, among others) can destroy value in the long term.

6.2.3 Framing an EOR Problem

Now we must answer the question, How do we frame the EOR problem to avoid the destruction of value? By the way, we discussed only one possible way to diminish value, but it can be done in many ways. Framing a decision problem can come in many flavors. We propose a scheme that is very similar to that in the 2001 Skinner book, *Decision and Risk Management for Reservoir Engineering*. The steps, as graphically represented in Figure 6.1, can be summarized as follows:

Policy. The idea of this step is that upon exchange of ideas between the decision support team or analysts (i.e., consultants) and the decision maker (it could be an organization), the main frame or constraints are established. It helps to define the conditions that limit the decision problem at hand. It includes aspects such as defining the core business, establishing environmental constraints for a given industry, putting limits on IRR and other financial inputs, land usage, waste disposal, and so on. The decision support team uses these constraints, but it does not attempt to introduce policies that create a conflict with them. These are boundaries to the decision problem, and they must be considerations of the design, but they should not be challenged or focused on. This first step, illustrated at the top of Figure 6.1, is the top of the so-called decision hierarchy.

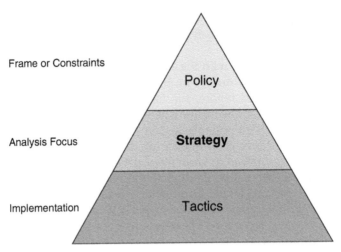

FIGURE 6.1   The decision hierarchy. *Source: Adapted from Skinner, 1999.*

Strategy. The core of the support team interest is this middle part of the pyramid. These are the issues that the team must deal with so an informed decision can be made. It is the place to create options or alternatives and to brainstorm and debate to build the true pivoting aspects of the decision problem. For EOR, you will create the alternatives through screening, review and analysis of field experiences (if any), lab studies, simulations, and other tools.

Tactics. This refers to the implementation aspects of the decision problem. The team should not be involved in it. For instance, the manner in which a contract for a chemical or gas supply is made is out of the context of the central strategy aspect. How the workforce that is necessary for the implementation is a tactic or strategic decision to be made by the organization (do not confuse this with the previous strategy).

The decision hierarchy may require brainstorming sessions, several meetings with the decision maker, and other strategies depending on your particular setting. It is vital that you do not rush this process until the focus of the issue has been clearly defined. Time and time again, many embark on a futile exercise of failed analysis, wasting resources on the wrong target. Interesting cases can be discussed, but once you have pinpointed the right problem, efficiency is gained and waste is diminished.

After your problem has been clearly stated—meaning that the structure of the problem has been established—you need to create the evaluation model. We strongly suggest the use of influence diagrams for communication purposes between consultants or support teams and decision makers. This is one of the uses of the decision model, although it is less significant for direct evaluation. Still, it might turn out differently for you. Some examples are provided for clarity. A second tool that is often used is the decision tree, which we describe along with influence diagrams, because the software we use allows one to generate decision trees automatically from influence diagrams.

6.2.4 Using Influence Diagrams and Decision Trees

What are influence diagrams and decision trees? Figure 6.2 illustrates the elements of an influence diagram. The different ovals, rectangles, and so on, have been labeled. These are the so-called nodes that, as you will see shortly, are connected by links. Each of them represents a *value*, a *discrete uncertainty*, a *continuous uncertainty*, and a *decision* that serve as elements of the decision model. For instance, an objective function such as NPV that is represented by a value node and construction should be located at the rightmost part of the diagram because these diagrams are constructed from right to left. Variables or influences can be represented on value nodes and depend on the type of influence diagram software you use (you could just this do in a presentation for

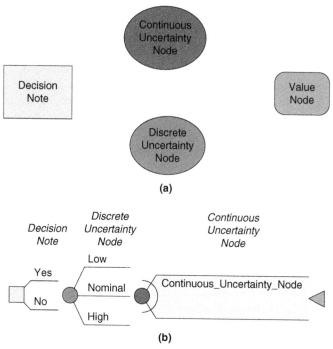

FIGURE 6.2 Unlinked schematic influence diagram constructs (a) and elements of the corresponding decision tree (b).

communication purposes). Vectors, arrays, or other quantities that are more complex than a single scalar can be defined.

Uncertainty nodes are used to represent quantities with a degree of uncertainty through either discrete probability sets or continuous probability values (or probability density functions). Let us give you a few words of advice. First, all quantities we measure contain a degree of uncertainty, but you should try to create uncertainties that are critical. What this means is that perhaps you know that pressure is a critical variable for an EOR process but brine salinity is not. There is a point to including an uncertainty node for pressure, but there is no real reason to include salinity as an uncertainty; instead, it should be included as a value node, if relevant. If you include every variable as an uncertainty, you will create an unmanageable evaluation model that will mask and deter the purpose of this exercise.

Second, ill-defined decision problems cannot be represented by uncertainties because a probability metric (measure) is used to deal with uncertainties, and this is not amenable to ambiguous treatment. Ambiguity should be entertained during the formulation of the problem (decision hierarchy, most likely) but not during evaluation. Some types of ambiguity can be treated as soft issues but not the other way around, meaning that soft issue analysis is not a general way of treating ambiguity.

Decision nodes are used to represent the options that the consulting or support team created. This decision node can contain a set as simple as "Yes" or "No" for deciding whether or not, for instance, a process should be moved forward. The decision node could be more complex and include a number of options—say a decision among a number of EOR processes or implementation options. You can manage to compare IOR options in general, such as well architectures, and EOR processes, but internal policies with the decision model might be necessary because the objective function may not be adequate for comparison purposes.

The bottom portion of Figure 6.2 shows elements of the associated decision tree. These trees, as opposed to influence diagrams, are built from left to right. The decision tree propagates value functions from left to right, included "risked" functions, for which probabilistic approaches are necessary. You should be able to frame and evaluate the decision model with either tool, but we find it easier to build the influence diagram because it provides a clear pictorial representation of the decision model.

Figure 6.3 shows the influence diagram with links corresponding to the initial set of nodes in Figure 6.2, as well as the corresponding decision tree. Notice that the links or arches (arrows) point to the right. The decision tree has not been completed because the influences represented in the diagram are qualitative—meaning that diagram shows what impacts (influences) the objective function may have (the rightmost

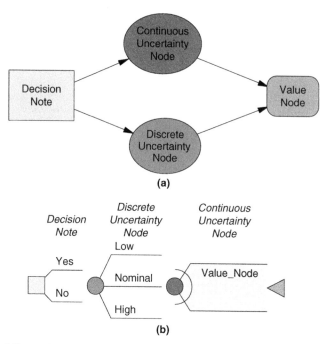

FIGURE 6.3 Linked influence diagram (a) corresponding to unlinked diagram in the figure and decision tree (b).

node) but not how they do so functionally. Once this is established, the decision tree can be completed, and the model can be run.

Figure 6.4 shows a simplified influence diagram for deciding which EOR processes should be applied. The objective function is the NPV, which implies that regardless of how much oil is recovered, the process with the largest NPV will be selected. The model contains no uncertainties. Notice that if the model were built to its next level, it would show cash inflow and outflow. We split the expenditure into OPerational Expenditure (OPEX) and CAPital Expenditure (CAPEX). OPEX accounts for operation costs, while CAPEX typically refers to infrastructures such as infill wells, treatment units, pumping systems, and the like.

For illustration purposes, we change the definition of two nodes to introduce uncertainties. As you probably know, oil price is uncertain. We redefined the value node as a discrete uncertainty node (Figure 6.5). We have also made oil production uncertain, but in this case we made the node a continuous uncertainty node. As you can see, the decision tree has changed, and the uncertainties will have to be risked. It turns out that because of the dependence on the uncertain inputs, the decision might change because now NPV will turn out to be a weighted average of several possible outcomes.

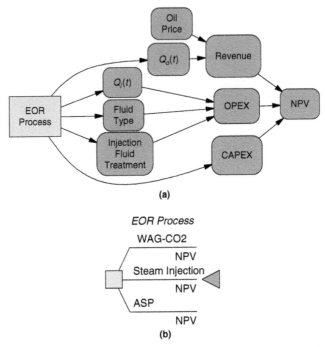

FIGURE 6.4 Simplified cash flow influence diagram for EOR (a). Simplified decision tree model without uncertainty (b).

6.3 ECONOMIC EVALUATIONS

In general, economic evaluations of enhanced oil recovery projects are company specific and usually developed in-house. Therefore, we will not overemphasize a specific methodology for EOR project economic evaluations. We would like to acknowledge the importance of economic analysis at different stages of an EOR project evaluation. Cunha (2007) published an interesting article addressing the importance of economics and risk analysis on today's petroleum engineering education, highlighting the importance of understanding the basic economic drivers in the oil and gas industry.

This is especially true when considering the screening and evaluating of EOR potential in a particular reservoir, independently of the experience of the engineers performing these tasks. It is common to see bright, smart engineers spending time and resources trying to solve and iterate problems without considering economics. Indeed, this situation is very familiar to this book's authors as a result of their participation in different organizations.

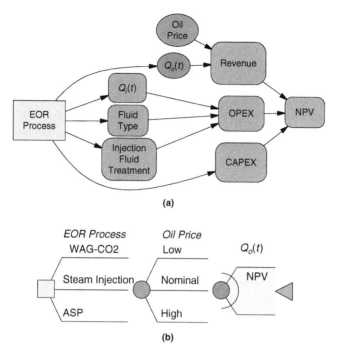

FIGURE 6.5 Influence diagram (a) and decision tree (b) with uncertainty for cases in Figure 6.4.

Both authors of this book come from a basic research-oriented education and more than once proposed what seemed to be great ideas but that lacked economic common sense. With time ("maturing"), we learned our lesson the hard way. No matter how promising a technical proposal may be, the final decision has to be scrutinized by using financial considerations. This is not to say that every step of an EOR evaluation has to be heavily weighted by economics, but it does mean that decisions that imply committing resources beyond an exploratory phase of ideas and concepts in EOR will come from decision makers, and at that point finances will become the final screen.

Therefore, in the proposed methodology economic evaluations are just simple estimations ("economic screening") that help evaluators to rapidly identify whether a particular EOR processes is technically and economically feasible enough to justify more in-depth ("expensive') and time-consuming studies (i.e., reservoir characterization, including field data gathering), special core analysis and fluid characterization, detailed numerical simulations, and surface facility evaluations required to handle produced fluids, among others.

It is well known that simple economic models and indicators are used to rank EOR options as part of the screening and decision analysis process. Operators or investors generally take over at this step, unless the supporting team and decision makers are part of the same organization and an internal decision has been made to let part of the same team carry on with the economic evaluation. The economic calculations can be linked to simulations and decision risk analysis either by exporting all possible production profiles to commercial software or by developing the required interfaces in Excel or Visual Basic. In these cases, the calculations are run by considering the main input criteria and constraints provided by the operator or investor. However, as described in Section 6.6, many feel that injection and production profiles should come from complex numerical simulation studies to be able to run the economics.

Although this is not entirely false, it does mean that economic evaluations based on simplistic models can be useful to rank and contribute to the decision-making process in EOR evaluations (Manrique et al., 2008). One good example is the economic evaluation of EOR methods based on production profiles generated in analytical simulations. EOR performance predictions using conceptual simulations generate optimistic production profiles ("most optimistic scenario") that can be combined with well costs, capital expenditures, operational expenditures, and various crude oil prices for the purpose of conducting preliminary economic evaluations. If projects are not profitable or are too sensitive under the conditions evaluated, they can be discarded and reexamined during the next planning period. However, later examination will depend on the risk tolerance of the operator or a particular investor for a specific EOR technology.

In other words, if a particular EOR project results in a poor economic performance using optimistic production profiles, further investments of time and resources are not justified. It is important to remind you that analytical simulations use idealized or optimum (optimistic) displacement efficiencies and cannot incorporate reservoir heterogeneities that are "always" present in oil reservoirs.

Therefore, when incorporating a more realistic geologic representation of the reservoir under evaluation, most cases will show worse economic performance than that estimated using optimistic production profiles generated in analytical simulations, unless a significant drive mechanism comes into play as a positive contributor in the numerical model. In this case, what one considers a limitation of analytical simulations (i.e., a too optimistic production profile) turns out to be an advantage for screening purposes and the ranking of EOR potential of multiple reservoirs in a short time frame.

6.3.1 Basic Economics of Steam Flooding

To provide a general idea of how basic economic indicators can be used in a simple way to estimate the economic feasibility of a particular EOR process, the following describes a field case scenario where an infill-drilling program is required to accelerate ("improve the rate of return of") oil production under steam flooding. This example consists of an extended steam injection pilot (multistaggered line-drive pattern) at well spacing of approximately 20 acres. Given the slow response shown during the pilot, a quick evaluation was developed to preliminarily identify optimum well spacing because a 3-D geological model was underway and no representative numerical model was available.

Basically, the operator was interested in understanding whether well spacing of 2.5 to 5 acres was required, based only on referential experiences from the available literature (Hanzlik and Mims, 2003; Keplinger, 1965; Ramage et al., 1987; Ziegler, 1987). Thus, a simple numerical model using general properties of the field under evaluation was used to answer the operator's question. Once the numerical model was built, a number of sensitivity analyses were run to evaluate the impact of well spacing on incremental oil recovery at different pore volumes of steam injected at 80 percent steam quality.

Figure 6.6 shows the incremental oil production as well spacing is reduced. Stages I through IV are representations of the incremental oil gained for every infill drilling program implemented in the steam flooding, maintaining the same well pattern configuration.

Intuitively, and assuming no economics is considered in this evaluation, an evaluator can propose a well spacing between 1.25 and 5 acres

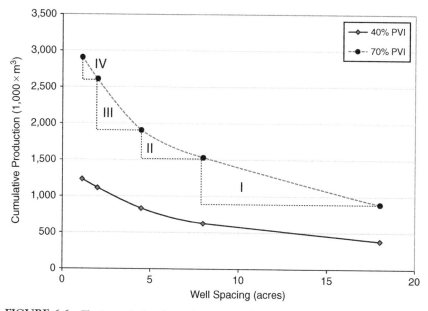

FIGURE 6.6   The impact of well spacing on cumulative oil production at two different pore volumes of steam injected.

based on the incremental oil obtained at lower well spacing, as shown in Figure 6.6. However, if the evaluator validates cost per well and considers the number of wells required for every infill-drilling program, he or she will have a general idea about the CAPEX associated with the wells in the scenarios evaluated. Knowing the level of investment required for drilling and completion of injectors and producers (and its flow lines, among others) for each stage, and estimating simple incomes from the oil produced at different oil prices, can contribute to a better interpretation of the scenarios evaluated.

Figure 6.7 shows the same results given in Figure 6.6, taking into consideration well cost (e.g., US$500,000 per well) and different oil prices (US$25 and US$50 per barrel). The results clearly show an inflexion point at a well spacing of 2 acres. In other words, the incremental oil recovery obtained at this well spacing does not offset the costs associated with the number of wells required (176 new wells), assuming the oil prices used in the evaluation.

Therefore, minimum well spacing might be around 5 acres after adding all associated costs, with steam injection projects not considered in this exercise. Of course, the trend can change when oil prices are higher, but this is part of a different risk and uncertainty analysis. The pore volume of the steam injected also provides a general idea of the timing at which crude oil can be produced. At lower well spacing, larger

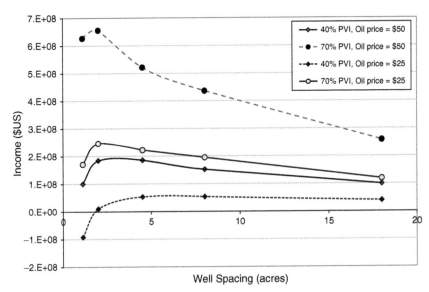

FIGURE 6.7 The impact of well cost (infill drilling) for the incremental oil production under steam injection shown in Figure 6.6. Income was estimated by the incremental oil (at two different oil prices) minus well cost ($500,000 per well).

pore volumes of steam can be injected in a shorter period of time (more steam injectors), accelerating incremental oil recoveries but incurring much larger CAPEX (e.g., wells, steam, natural gas, water treatment, and handling).

Additionally, when considering infill-drilling programs, it is important to understand how long the drilling and completion of one well take and the number of drilling rigs available. This information will probably provide a quick estimate as to whether the project is feasible in terms of field implementation. We believe this approach is very useful for engineers who do not have a strong background in economics because engineers can intuitively recognize whether a particular EOR project (steam flooding in this case) is technically and economically feasible. The latter does not mean that a rigorous economic evaluation cannot be done based on simulation results and all costs associated with steam injection projects. However, the main idea of this approach is to provide basic tools to young engineers and to advise them to always remember the importance of economics in EOR evaluations.

6.3.2 Basic Economics of Chemical Flooding

Infill drilling has become an important consideration in the evaluation of EOR chemical flooding in mature waterflooded and heavy-crude oil reservoirs for the following reasons:

- In recent evaluations, the authors faced the situation of having to analyze EOR chemical flooding (SP or ASP) in waterflooded reservoirs that require infill drilling to accelerate oil recovery and improve economic indicators (e.g., return on investment). In other words, EOR chemical flooding not only has to be profitable in itself, but also has to pay for infill-drilling programs, making the implementation of this type of project difficult. The reasons that infill drilling cannot be justified are diverse; some examples include highly mature waterfloods (water cuts >95%), lack of reservoir characterization, and high uncertainties associated with remaining oil saturations, among others.
- Regarding EOR chemical flooding (AP, SP, or ASP) in viscous oil reservoirs, a common question that arises is the potential injectivity reduction (polymer injectivity) due to injection of a more viscous injection fluid regardless of reservoir permeability (K) or flow capacity (k^*h), where h is the net thickness of the reservoir. Adding new vertical or horizontal injectors can mitigate potential injectivity reduction, especially in shallow reservoirs where well costs are not too high. However (and as explained earlier), CAPEX associated with drilling and completing wells may negatively impact project economics.

Therefore, and given the complexities of EOR evaluation, variables such as well spacing and injection volumes had to be evaluated as early as possible during the screening phases to validate the economic feasibility of a particular EOR method (Teletzke et al., 2008). Keeping in mind the importance of injectant costs (e.g., cost of imports and transportation, including taxes) during the early stages of screening field candidates for EOR methods will help to prepare stronger cases and to facilitate the decision-making process at the management or investor level.

If as an EOR evaluator you prove that a particular case shows enough economic merit, the project can be funded and potentially reach a pilot project phase to validate the real potential of the proposed EOR method even at low oil prices. If the concept proves successful in a pilot study (or a lab and simulation), your company will be ready to deploy a larger (or pilot) application when oil prices reach an adequate comfort level for the decision maker, timing being part of the risks and rewards of EOR applications.

However, if oil prices reach a comfort level after several years of the pilot, do not forget to check if the products evaluated at the lab scale or injected in the pilot are still commercially available. It is possible that some chemical reagents will no longer be available at the commercial scale, which may cause delays and even the cancellation of pilot projects. Therefore, it is important to indicate that oil reservoirs need to be reevaluated periodically to determine not only changes in reservoir conditions (e.g., changes in reservoir conditions with time) but also changes related to a particular technology, chain of supply, and the like.

This demonstrates the complexity of EOR implementation; moreover, every single application is specific to each reservoir. Again, for those without a background in economics or who do not have direct support from economic evaluators, it is important to remember the importance of costs during your evaluations and for a successful application of EOR technologies. Detailed economic evaluations are not always required to recognize if a particular EOR process is technically but, most important, economically feasible in the field under evaluation. EOR economics can be impacted by special incentives and tax regulations. For this reason, the next two sections discuss the importance of costs in CO_2-EOR storage projects and the impact of special incentives and taxes on EOR implementation.

6.4 CO_2-EOR STORAGE AS IT RELATES TO COSTS

CO_2-EOR storage has already been discussed, so this section briefly covers how the costs of different phases of carbon capture storage (CCS) projects impact their economical feasibility. Uncertainties associated with the economics and regulatory framework of CCS projects have been widely addressed in the literature because they have been identified as the main limiting factors for further development of such projects in the United States and abroad (Algharaib and Al-Soof, 2008; Gaspar et al., 2005; Ghomian et al., 2008; Hustad, 2009; Imbus et al., 2006; Negrescu, 2008, Xie and Economides, 2009).

The economics (costs and revenues) involved in a CCS project can be broken down in many ways. However, it depends on the source of CO$_2$ (e.g., petrochemical plants vs. coal-fired power plants) and where it will be injected (e.g., EOR vs. saline aquifer). Assuming a scenario of CO$_2$ capture from a coal-fired power plant, we can divide the main economic variables into four categories: CO$_2$ capture and compression, CO$_2$ transportation, CO$_2$ storage (including wells and monitoring), and possible revenues (e.g., oil recovery and/or carbon credits), depending on the application (CO$_2$-EOR vs. saline aquifers).

All four areas are considered complex by nature and require a high level of integration in terms of evaluation and implementation with CO$_2$ storage projects. However, CO$_2$ capture and compression represent the most expensive phases of these projects (up to 80 percent) assuming today's commercially available technologies. CO$_2$ capture costs usually include compression costs as well as the cost of energy consumed during this process once the CO$_2$ has been separated.

The energy requirements of CO$_2$ compression represent an important energy penalty for coal-fired power plants, reducing the net output of the plant at capacity and therefore requiring power replacement and the subsequent increase in electricity costs. Finally, CO$_2$ compression and transportation costs also depend on, among other factors, the physical distance of the power plant to the pipeline and the CO$_2$ flow rate attainable in the pipeline (Manrique and Araya, 2008).

CO$_2$-EOR storage has become the preferred emission reduction strategy because incremental oil production can offset, if not exceed, all of the costs associated with funding such projects until a proper regulatory framework is in place. Figure 6.8 shows an example of the main costs and revenues associated with CO$_2$-EOR storage.

Costs of the main categories shown in Figure 6.8 come from a comprehensive literature review done by Manrique (2008) for a private company where the CO$_2$ capture cost included the costs of CO$_2$ compression. Oil, carbon credits, and rates of exchange were obtained from different Internet sources (MSN Energy, 2009; Point Carbon, 2009a; *x-rates.com*, 2009).

Regarding the carbon credits, the use of EUA OTC (European Union Allowances–Over the Counter Market) was arbitrarily selected because of easy access and the possibility of tracking historic prices as well as market definitions (Point Carbon, 2009b). From Figure 6.8, an approximate cost of US$100 per ton of CO$_2$ was estimated based on our reviews

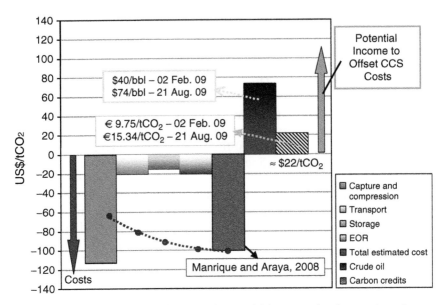

FIGURE 6.8 Example of total costs and potential incomes of carbon capture storage projects with EOR (CO_2-EOR storage).

(total estimated cost per ton of CO_2 captured and stored), while the incomes (oil prices and carbon credits) show their volatility in two, randomly selected months during 2009. For the carbon credits, a rate of exchange of US$1.43 per euro was used (*x-rates.com*, 2009).

It is important to mention that in April 2008, the average oil price was US$119/bbl and EUA OTC was reported to be at US$38.05/t$CO_2$ (Manrique, 2008). Therefore, combining price volatility of oil prices and carbon credits with costs associated with CO_2 capture, compression, and transportation (if pipelines are not readily available), CO_2-EOR and storage might not be economically feasible unless additional incentives are developed. However, some incentives proposed for CO_2-EOR using anthropogenic sources of CO_2 are addressed in the next section.

On the other hand, carbon credits will definitely not offset costs associated with CCS in saline aquifers (without EOR), at least in the near future (Figure 6.9). Additionally, the cost of CO_2 storage in saline aquifers might be too high because of the lack of reservoir characterization and existing infrastructures (wells and surface facilities) compared to oilfields. As mentioned earlier, CO_2 capture and compression represent a capital-intensive phase of any CCS project. Therefore, it is believed that with the refinement of current capture technologies and/or the development of new technologies, the opportunity will exist for major cost reductions in CCS projects (see Figure 6.9). These cost reductions may

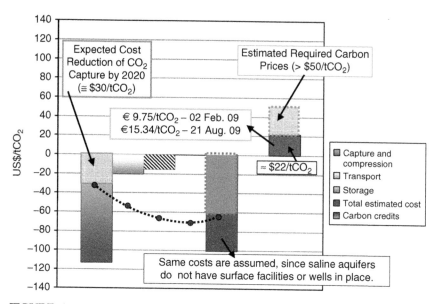

FIGURE 6.9 Example of total costs and potential incomes of carbon capture storage projects without EOR (CO_2 storage).

contribute to future CO_2 storage project economics. CO_2 capture costs below \$30/ton are considered necessary in order to justify the economic feasibility of CCS projects in the near term, assuming that the proper regulatory environment and carbon market framework are in place.

Ultimately, governments, the oil and gas industry, and utility providers will need to integrate multidisciplinary efforts in order to identify optimum CO_2 storage options. Future development will continue to define carbon trading markets and the necessary regulatory framework. It is important to note, however, that economic and legal issues are not the only challenges facing CO_2 storage projects. Other issues include identifying geologic sinks that can safely store CO_2 and conducting community-based initiatives to educate the general public on the economics and safety of CO_2 capture and disposal, especially with regard to power generation plants.

6.5 EOR INCENTIVES

For decades, several countries and some U.S. states have proposed, considered, or adopted tax incentives designed to improve the economics of marginal and mature fields. One of the major concerns of operators and governments is the daunting number of reserves that could be abandoned because EOR potential has yet to materialize as reservoirs approach maturity and because of the high volatility of energy prices. Therefore, the main objective of these incentives is to recover additional reserves and extend the productive life of the reservoirs by implementing EOR methods. Numerous references address these taxes and incentives, including Brashear (1994); Brashear et al. (1989, 1991); Buckner (1994); Griffith and Cox (1986); Henry (1977) Iledare (2004); Jayasekera and Goodyear (2002); Sharp (1975); Terzian et al. (1995); and Thompson and Wright (1984).

EOR incentives are specific to country and state (in the case of the United States), so this is not a comprehensive discussion of exemptions, limits, and rates. Instead, we hope we have given you some information that you will find useful when evaluating EOR potential in a particular field, especially in the United States. Therefore, reviewing possible EOR incentives in states or countries is strongly advised. The following cases are good examples of some incentives that have been proposed, considered, or adopted in the United States and abroad:

- The U.S. National Energy Strategy considers a tax credit for specific investments and injectant costs for qualified EOR projects (IRC §43 EOR credit). Several rules and conditions apply to the EOR tax credit; one of them is oil price. The credit is 15 percent of the qualified EOR costs paid or incurred by a taxpayer in connection with a qualified project. In 2006 the credit was phased out because of high oil prices. However, and based on oil prices, a notice is issued stating whether the credit is available (Buckner, 1994; IRS, 2007).
- The state of Texas provides a special EOR tax for approved new EOR projects or for the expansion of existing ones. The rate is 2.3 percent of the production market value (one-half of the standard rate) for 10 years after Commission certification of the production response. In 2007, Texas also approved an "Advanced Clean Energy—EOR Tax Reduction" for CO_2-EOR storage projects using anthropogenic sources (RRC, 2007).
- To promote the implementation of EOR projects, in 2003 Wyoming reactivated a reduction in severance tax from 6 to 4 percent on the first five years of incremental production coming from EOR operations (Nummedal et al., 2003).
- Saskatchewan Energy and Resources has developed a CO_2-EOR incentive plan offering cost share support over five years ($7.2 million) to promote the design and implementation of CO_2-EOR pilot projects. Additionally, the Saskatchewan Petroleum Research Initiative (SPRI) offers up to $3 million in royalty/tax credits per project. The main objective of this initiative is to demonstrate technical

and economical feasibility and to increase awareness of CO_2-EOR projects (Saskatchewan Energy and Resources, 2009).

- Hustad and Austell (2004) propose the joint participation of governments of adjacent countries in the North Sea and commercial institutions to develop mechanisms that create incentives for the oil industry to obtain CO_2 from anthropogenic sources for CO_2-EOR projects. This will encourage CO_2 emitters to invest in capture technologies that can be used for CO_2-EOR and thus extend the production lifespan of existing fields for one or two decades.
- By the end of 2008, Argentina had approved the Programa Petróleo Plus, or PPP (Oil Plus Program), as a strategy to increase oil production and oil reserves. Argentina is facing a steep production decline, and PPP will allow oil producers to sell incremental oil beyond the US$47/barrel regulated price. The program is based on incentives or tax returns, or on a combination of the two that may give oil producers an estimate from US$7 to 10 per incremental barrel. EOR is one of the areas that can apply for the program (Crooks, 2008; Rossi, 2008).

The preceding discusses just some of the ongoing and proposed initiatives that may contribute to the implementation and the potential expansion of EOR projects during the next decade or two. This chapter addressed a number of economic factors that need to be considered during evaluation of enhanced oil recovery to avoid the abandonment of reserves. The authors believe that governments and operators need to work together to increase reserves and energy security, delaying field abandonment for as long as possible. If they do, EOR may still have a potential that has never been tapped.

Methodology

7.1 INTRODUCTION

This is the pivotal chapter of this book. Here, we combine the elements we discussed in detail in the preceding chapters and sections to integrate the strategy that is conducive to enhanced oil recovery (EOR) projects. Published results show that a combination of conventional and advanced EOR screening with fast analytical or small-scale numerical simulations represents a valuable approach to property acquisition and evaluation to support reservoir development plans (RDPs) and, most recently, to identify EOR opportunities in connection with carbon storage opportunities. To illustrate the different types of decisions, contexts, and constraints of the decision-making process, cases are divided according to the availability of data and the time constraints for the decision-making process.

We divide the cases into two groups. Field case type I is characterized by a limited amount of data and a relatively short time frame for making decisions. This type of decision-making problem emphasizes the screening steps rather than the entire workflow because, as Figure 7.1 shows,

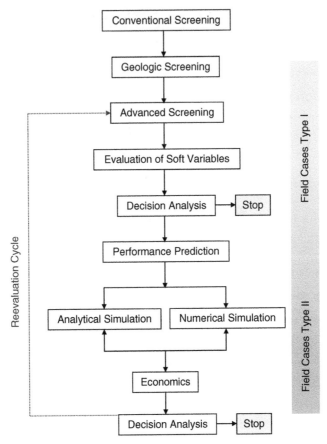

FIGURE 7.1   EOR decision-making workflow. Each field case type emphasizes a portion of the workflow.

the decisions for this type of asset are often framed in terms of data-gathering initiatives or they focus on aspects of feasibility.

In contrast, type II field cases are not limited by the amount of necessary data but mostly by the time span allotted to the decision. This condition allows a focus on performance prediction, so the effective use of simulation tools is a must. You can think of the flowchart (see Figure 7.1) as a sequence of qualitatively different screening stages. Although the amount or type of data representation at each stage is different with each increased level of complexity, these stages do not necessarily represent a hierarchy; instead, different representations of the reservoir–field systems are adopted. The field cases described in this chapter illustrate the two types of decision-making problem. The steps in the proposed methodology are described in the next sections.

7.1.1 Conventional Screening

In this phase, the first of six steps, engineers and analysts can determine a lack of published field experience for those methods in fields that have reservoir properties similar to the one under evaluation. This idea is the basis of reservoir analogs, on which we will elaborate more extensively. The choice of variables is guided by both the availability of data and the analyst's "intuition." Radar plots of up to six variables are also used to identify trends and ranges of preferences for (or the applicability of) a particular enhanced oil recovery method in multiple reservoirs prior to beginning advanced screening (Manrique and Pereira, 2007). In the case of commercially unproven EOR processes (THAI, VAPEX, etc.), this phase helps to analyze and estimate the technical feasibility of these recovery methods based on a more theoretical and engineering basis. The notions conveyed by comparing the reservoir–field problem at hand with abundant recorded field cases are the level of risk and bias containment so that what is considered "wisdom" does not become an excessive cognitive bias.

Conventional screening is complemented by the use of screening options in commercial analytical tools to expand the evaluation and thus the further validation of the applicability of the most feasible recovery process for the field that is under evaluation. This analytical screening is also based on the comparison of the field's reservoir properties with property intervals that are derived from analyzing known EOR projects in the databases. Two known procedures are used to expand on screening.

The first procedure is based on "go–no go" criteria (ARC, 2006), while the second one uses fuzzy logic to generate scores for ranking based on a triangular distribution of comfort intervals (IRIS, 2007). Since the "bias" (expertise bias) differs in the two screening procedures, these additional approaches provide a comprehensive evaluation of the property of interest. This form of screening can be considered the reservoir engineer's typical focus of attention—namely, average reservoir and fluid variables.

7.1.2 Geologic Screening

Geological characteristics, such as trap type, depositional environment, lithology, structure type, and digenesis, are used to compare EOR projects in a database with the field under evaluation. This representation is intended as a simplified geologic view to cluster fields into reservoir types. For sandstone reservoirs, this analysis is augmented by use of the matrix of depositional environment versus lateral and vertical heterogeneities (Henson, Todd, and Corbett, 2002; Tyler and Finley, 1991).

The classification can be somewhat subjective because of a lack of geologic information and/or differences in geologic interpretation, but it can still guide the EOR decision-making process using field experiences. If the dimensions of sand bodies or genetic units (length, thickness, and width) and current or proposed well length and spacing are known, horizontal and vertical heterogeneities indexes can be estimated through simple equations (Henson, Todd, and Corbett, 2002).

Additionally, 2-D and 3-D heterogeneity index analyses have been shown to be more robust when combined with the Dykstra-Parsons (DP) coefficients calculated from well log (petrophysical analysis) and core permeability data. DP coefficients have also been used extensively to generate full-field maps as a quick quality-control procedure during full-field petrophysical studies—that is, the impact of petrophysical cutoffs on reservoir heterogeneity as a part of detailed reservoir ("integrated") engineering studies. DP maps combined with other reservoir properties—for example, net pays and fluid saturation—have also been used in full-field analytical simulations to evaluate RDP under EOR processes (Alvarado et al., 2003).

7.1.3 Advanced Screening

Advanced screening is purposely designed to aggregate field cases into reservoir typologies. This task is generally complex, using simpler 2-D representations because the true analogy is multidimensional. The techniques used, which are based on artificial intelligence, data mining, and space-reduction, are well documented in the literature (Alvarado et al., 2002; Manrique, Ranson, and Alvarado, 2003). To perform this advanced screening, mined data from roughly 450 successful EOR projects are compared with the reservoir(s) under evaluation. The simultaneous projection of a reduced set of reservoir variables—namely, temperature, reservoir depth, current reservoir pressure, porosity, permeability, API gravity, and viscosity—is represented on 2-D maps (expert maps). Clusters in these 2-D projections represent different reservoir types (reservoir typology).

Experience shows that 2-D representations of reservoir clusters share the types of EOR project implemented. Multidimensional projections on the 2-D plots offer simultaneous comparisons of multiple variables and, more important, a convenient clustering of reservoir types. This is how statistics on recovery factors can be obtained, adding robustness and reducing those "expert opinions" in the screening evaluation.

This screening technique was developed specifically for CO_2 injection for CO_2-EOR and sequestration evaluations (Manrique, Ranson, and

Alvarado, 2003; Velasquez, Rey, and Manrique, 2006). As indicated earlier, some aspects of screening cannot be quantified.

7.2 EVALUATING SOFT ISSUES

Evaluating soft issues early in the procedure can prevent time wasted considering EOR processes that are not feasible—for instance, because of lack of injection fluids or offshore restrictions. Some of these decisions involve ongoing technology developments that are not yet entirely proven. Such cases are not amenable to the same type of "hard" analysis based on economic indicators.

7.3 PREDICTING PERFORMANCE

At this point, we should address not whether a process is feasible but whether, once implemented, it will have the potential of incrementing production or a recovery factor. Based on the decision framework, predicting performance using analytical and numerical simulations or both is defined at early stages of the evaluation. Numerical simulation studies are costly and time consuming in addition to requiring highly trained professionals. In some cases, full numerical reservoir simulation studies are not justified because of the lack of available data and/or time constraints.

It is true that oil production forecasts obtained from analytical simulations tend to be overly optimistic, given their limitations. However, for fast screening purposes, analytical simulations provide key insights, sensible parameters, and a way to identify the uncertainties associated with different recovery processes. If projects do not offer economic merits using the optimistic production profiles in analytical simulations, most certainly the project economics will be less attractive when more detailed simulation studies are completed. In the case of numerical reservoir simulations, conceptual or sector models, instead of the full-field model, can be used to complete this step.

7.4 ECONOMICS

Simple economic models and indicators are used to rank EOR options as part of the screening and decision analysis process. Operators or investors generally take over this step, unless the supporting team and decision makers are part of the same organization. The economic calculations can be linked to simulations and decision risk analysis either by exporting all of the possible production profiles to commercial software or by developing the required interfaces in Excel or Visual Basic. In these cases, economics is run by considering the main input criteria and constraints provided by the operator or investor.

The economic evaluation can also be run using optimistic production profiles and different well costs and oil prices for screening. If projects are not profitable or are too sensitive under the conditions evaluated, they can be discarded and reexamined during the next planning period. However, later examination will depend on the risk tolerance of the operator or a particular investor for a specific EOR technology.

It is important to note that an oil reservoir must be reevaluated periodically to determine how current development plans may impact EOR processes in the later stages of production because of changes in reservoir conditions. Potential oil recovery is dynamic, changing as a reservoir matures and as its energy evolves. The next section describes six field cases that were evaluated, in the United States and abroad, as well as the decisions made based on the results of this fast screening analysis and evaluation.

7.5 FIELD CASES

The six cases presented here are meant to illustrate a variety of decision-support exercises. All of them refer to actual business decisions in the context of EOR, but field names and locations are not disclosed to protect the privacy of the parties involved. These field cases include shallow (500 feet) to deep (15,000 feet) reservoirs, with oil gravity ranging from 7°API (Canadian oil sands) to 60°API (South American gas condensate reservoirs). Field lithologies include sandstone (consolidated and unconsolidated) and carbonate reservoirs.

7.5.1 Field Case Type I: Time Constraints and Lack of Data

Enhanced oil recovery screening studies that lack data and have stringent time constraints (typically from a few weeks to up to three months) for the decision-making process are the most common decision-support cases developed in recent years that follow the workflow in this book. In terms of the steps of our methodology, the case studies cover mostly steps 1 through 5. However, the details of each screening study vary as a function of the information available and the decisions based on the results obtained in this type of analysis. Common decisions or questions that need to be answered from EOR screening studies can be exemplified with the following list:

- Determine the most feasible EOR processes, including preliminary analytical simulations to estimate oil recovery potential.
- Justify data-gathering programs: drilling and logging wells, core and fluid samples recovery, and so on.
- Justify detailed laboratory studies such as chemical waterflooding, minimum miscibility tests, special core analysis, and so on.
- Justify more detailed engineering (Phase II) studies.
- Generate preliminary RDP based on one or more EOR process, among others.

Case Study, Field A

The Field A case was posed as a screening problem (a light-oil dolomitic formation in the United States), including preliminary analytical simulations. The main objective of this screening study, developed in early 2007, was to identify the most feasible recovery processes and potential closer analogs to define a more detailed simulation study and potentially a pilot test. Results from the screening study show gas injection (continuous injection or in WAG mode) and waterflooding to be the most viable EOR options for this field.

Several reservoir analogs in which N_2 (e.g., Binger Field, Oklahoma) or CO_2 (e.g., Charlson, Kutler, and University Wadell) had been injected were found in the literature (Figure 7.2). Additionally, waterflooding projects in low-permeability (< 1.5 md) dolomite reservoirs were identified in some Texas fields (e.g., Levelland, Mabee, and Robertson North) after several queries to the database. Potential field ("closer") analogs identified helped to increase the operators' level of confidence associated with the field's development.

Although preliminary analytical performance predictions showed that gas flooding (continuous or in WAG mode) outperformed waterflooding, water injectivity tests were recommended to validate a WAG injection strategy because of the low average permeability of the field. Given the applicability of N_2 and CO_2 ("flue gas") flooding in Field A, high-pressure air injection (HPAI) was not discarded at this level of evaluation.

HPAI was considered an option after analyzing the availability of CO_2 (anthropogenic and natural) sources near the field and the estimated N_2 MMP (from empirical correlations), as well as the costs associated with air separation and N_2 rejection units. The study's results justified a data-gathering program to develop the required laboratory and field data for use in identifying potential pilot areas in the field for the evaluation's next phase.

Case Study, Field B

The Field B case included screening and evaluation of a Canadian oil sand property (Lloydminster and Rex formations). The main task in was

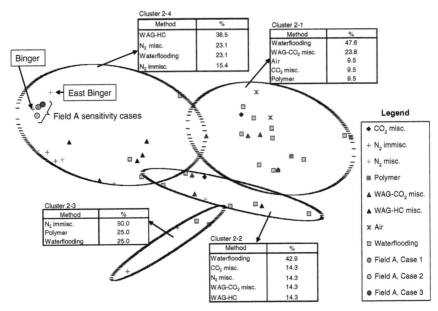

FIGURE 7.2 Expert map for Field A case advanced screening.

to identify whether the property could be developed by steam-assisted–gravity-drainage (SAGD) given unsuccessful cyclic steam injection pilot tests in the late 1980s. The study was divided into two phases: The first was screening, including preliminary analytical simulations; the second consisted of a 2-D simulation using an existing geologic interpretation of the property under evaluation. The methodology followed for this field case and several other Canadian oil sands was recently published (Manrique and Pereira, 2007).

The results of this screening study showed that both SAGD and steam injection (cyclic or continuous steam) were applicable to Field B. Potential field analogs were identified for SAGD (Burnt Lake) and cyclic steam injection (Cold Lake and Peace River). However, analytical simulations consistently showed that SAGD outperformed cyclic steam injection in a broad range of reservoir and steam injection conditions. Once the geologic interpretation became available, we noticed that Field B showed a lack of continuous net pay areas of 20 m or thicker, limiting the applicability of SAGD based on conventional screening criteria (see ARC, 2006; Butler, 1991; Palmgreen and Renard 1995).

Additionally, the geologic interpretation showed that thicker areas ($\geq$20 m) were not always continuous because of presence of interbedded thin layers of low permeability that might negatively impact steam chamber development and thus cumulative steam–oil ratios (CSOR) and oil recovery factors. To validate our findings, a 2-D parametric numerical simulation was carried out to estimate the potential of SAGD in Field B. In the absence of sufficient reservoir data to run a proper numerical study, PVT and relative permeability data were estimated from public documents. Sensitivities on net pay and the presence of top gas and bottom water were also run to estimate the impact of these critical variables on cumulative oil production and CSOR.

The 2-D parametric numerical simulation study helped to identify areas where SAGD was applicable. Additionally, detailed review of successful cyclic steam injection projects reported in Cold Lake, Peace River, and Primrose justified close study of past experience with this recovery method in the field. Finally, the screening study identified documented potential development schemes under SAGD and cyclic steam injection that are currently under consideration by the owner of the field.

Case Study, Field C

The decision problem associated with the Field C case was the well spacing for a steam flood project with a short time frame for decision making. Field C consists of a giant (2-billion-bbl) heavy-oil (19° API) field in Asia. The analysis was framed as a decision problem having efficient recovery–production as the objective function. Since building a representative geological and petrophysical model for numerical simulation was not feasible given the time constraints and available data, the problem was how to use the best of the available data to provide a meaningful recommendation in the given period of time.

In situations in which large volumes of data are available, but not necessarily sufficient, it may be tempting to argue that a detailed model will provide the solution. However, in this case the time was insufficient for a detailed numerical simulation, and key data to complete a detailed reservoir model were not available. Our approach was to integrate the available data and use analytical tools and some small conceptual models. This is not unlike the moving-mosaic techniques used for infill drilling (e.g., Guan et al., 2002; Hudson, Jochen, and Jochen, 2000; Hudson, Jochen, and Spivey, 2001; Voneiff and Cipolla, 1996; Wozinak, Wing, and Schrider, 1997) or the quality map method (e.g., Da Cruz, Horne, and Deutsch, 2004).

A frequent approach to analytical simulation consists of modeling a well pattern (generally 5-spot patterns) with the assumption that the obtained results represent the average performance of a particular sector of the reservoir. Given that this assumption is strong, as properties in a reservoir should be statistically equivalent to be valid, care should be taken in interpreting this type of evaluation.

For this purpose, conceptual or analytical modeling accounts for variability in the reservoir by evaluating the performance of each pattern with areal distribution on the reservoir map on a statistically equivalent basis and with the distribution of critical parameters for oil recovery processes such as the Dykstra-Parsons coefficient or some other heterogeneity index. The DP coefficient was used in Field C as a heterogeneity index to generate the quality map for each reservoir, as shown in Figure 7.3. Data from 384 wells were used to generate DP coefficient maps.

Several quality maps were obtained per reservoir—for example, the DP or net pay maps or a combination of both indicators. Although some other parameter could have been used to build a quality map, DP and net pay maps were the only ones used here. Multiple analytical simulations were run for each quality map. For example, the quality map shown in Figure 7.3 is divided into different heterogeneity regions (based on DP differences), and for each heterogeneity region several analytical simulation runs were completed, accounting for well spacing and samples of the available data. The recovery factor for each pattern for different well spacing was calculated, and the values of the recovery factor were then weighted by the value of the original oil in place.

The recovery factor is shown in Figure 7.4 for different well spacing values. Besides well spacing several other sensitivity parameters, such as steam quality, were studied. Figure 7.5 compares the recovery factor for different steam quality and well spacing after injection of 0.7 pore volume of steam.

The production and injection forecasts were transferred to a simple economic model to calculate the net present value for each scenario. The economic model was extremely simple. For this reason, rather than

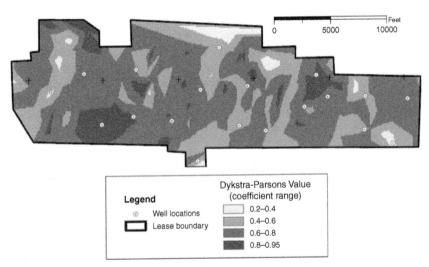

FIGURE 7.3 The Dykstra-Parsons coefficient map for one of the reservoirs in Field C.

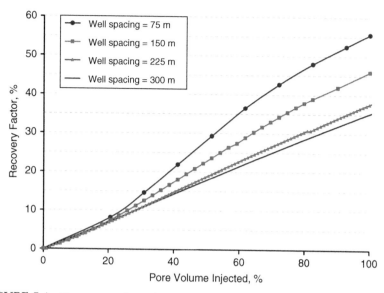

FIGURE 7.4 The recovery factor for one pattern among different well spacings.

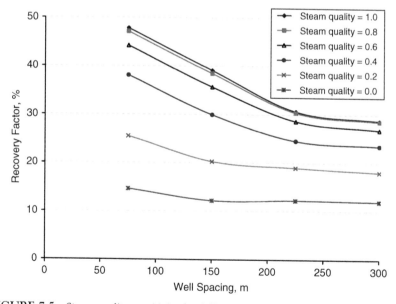

FIGURE 7.5 Steam quality sensitivity for different well spacings.

focusing on the accuracy of the NPV estimate, attention was centered on the relative economic performance of the different scenarios. Two oil price cases ($60/bbl and $90/bbl) were run. In any infill case a number

of wells had to be drilled. The drilling and completion costs were estimated at $1 million per well for both injectors and producers. This cost
included surface facilities, and the cost of steam was assumed to be
$1 per barrel. Net cash flow was calculated, with the results depicted
in Figure 7.6. These results show that the optimum well spacing of
approximately 150 m.

One interesting question is how much the accuracy of a model affects
the outcome of decision making. The answer depends on the project context, but some decision makers often distrust simplified modeling strategies because of their lack of accuracy. This is the case with analytical
simulation and conceptual sector models for EOR. Given the limitations
and uncertainties of data sources (models) and given time constraints,
simplified analyses are often a better solution to certain decision-making
problems (Bos, 2005). Embarking on EOR projects with a hierarchy of
successive interdependent decisions can facilitate the promotion of further development in a field.

Analytical simulations relied on layer-cake models, for which petrophysical data from interpreted well logs were used to assigned DP coefficients. This representation of heterogeneity added to the limitations of
analytical simulation can, in principle, be considered worthless from a
decision-making point of view. A comparison was drawn with numerical simulation models. The distribution of permeability in the numerical
model reproduced the DP coefficient's value, but unlike in the analytical

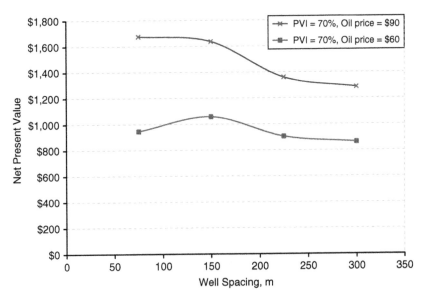

FIGURE 7.6   Net present value at 70 percent of pore volume injected with two different
oil prices.

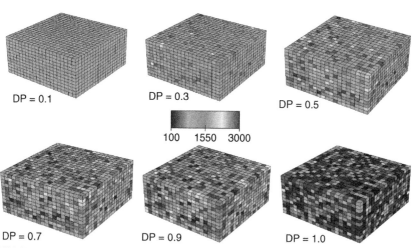

FIGURE 7.7 Permeability distributions for numerical models for Field C.

models, permeability was distributed isotropically in 3-D (Figure 7.7). Besides the difference in the solution procedure, the model in this case offered more degrees of freedom than the analytical cases.

The comparisons between numerical and analytical models are presented in terms of recovery factor versus percentage of pore volume injected. Steam flooding results are presented in Figure 7.8 for two of the well-spacing values considered. In steam flooding, higher well spacing results in higher heat losses; therefore for the same pore volume injected, the recovery factor drops in the higher well spacing. Numerical models show that the results from analytical models for different well spacing are optimistic.

Despite significant differences in production profiles (represented here by the value of the recovery factor), a postmortem analysis using numerical simulation results showed that the final recommendation was not affected by the simulation procedure. Although this will not always be the case, what is true is that even if a complete and sophisticated model is available, the reality of production can be badly predicted. Moreover, if time constraints are stringent, more complex models tend to yield suboptimal decisions (Bos, 2005).

7.5.2 Field Case Type II: Insufficient Time to Use Data

In this type of field case, the volume of data may be sufficient to build detailed models for which all levels of screening can be completed, but stringent time constraints (from two to four months) make EOR screening studies more challenging because the decision-making process is

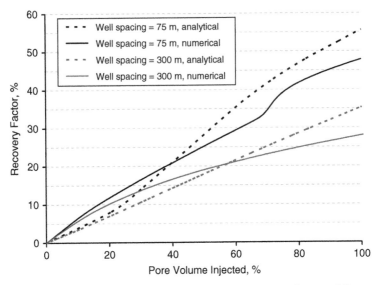

FIGURE 7.8   Comparison of analytical and numerical models for two different well spacings.

more complex. In these field studies, all of the steps of the proposed methodology can be covered. However, the changes and challenges are related to the simulation approach used to help the investor–operator make a decision.

As we said earlier, each screening study is specific, depending on available information, geographic location, access to oil and gas markets, and investor–operator decision frameworks, among other factors. The following are some of the questions that were answered as a result of EOR screening studies:

- Initiate visualization to estimate the most feasible recovery processes and RPD as part of front-end loading (FEL) studies.
- Evaluate EOR technologies and potential implementation strategies for one reservoir or a portfolio.
- Analyze reservoir portfolios to evaluate CO_2-EOR and storage potential.
- Justify large investment decisions to develop pilot tests or comprehensive data-gathering programs associated with EOR projects.
- Justify more detailed engineering studies (Phases III through V, as shown in Figure 9.1)—that is, full-field simulations studies, EOR project design and monitoring, investments in surface facilities, and so on.
- Complete property evaluation and acquisitions.

Some examples of this type of decision making, which was developed in countries in the Western Hemisphere using the proposed methodology, are briefly described next. Again, field names and locations are not disclosed for privacy reasons.

Case Study, Field D

The Field D case was part of a visualization study to identify potential EOR technologies and RPD in a portfolio of gas condensate and light–crude oil reservoirs (11 multireservoir fields) in South America. The identification of technologies, recovery processes, and production strategies to maximize gas and condensate production represented a key objective of the analysis. Ranking the most feasible options to maximize gas and condensate recovery was also a key objective of this analysis phase.

After evaluation of the hydrocarbon gas in place, stochastic volumes (P10, P50, and P90) were used to generate production profiles and cumulative recoveries considering different production scenarios. Gas and condensate production predictions were based on analytical simulations (development planning Excel spreadsheet). The model used assumes that the gas field obeys a relationship of pressure compressibility factor ratio (P/Z) versus cumulative gas produced. Despite the simplicity of the model, it is a rather rigorous program that includes drilling, inflow performance gas injection and/or reinjection, and surface compression, among other activities. Typical production profiles calculated from this evaluation and used in the economic calculations are shown in Figure 7.9.

Figure 7.10 shows the cumulative gas and condensate production for all fields for three different development scenarios: N_2 injection, gas reinjection in selected fields (selective gas reinjection) plus compression, and selective gas reinjection plus compression and coproduction.

Regarding EOR opportunities for this portfolio, N_2 injection was found to be the optimum strategy to increase gas and especially condensate recovery. Additionally, N_2 injection would contribute to maximizing gas sales by reducing dry gas reinjection. However, gas recovery would not be as efficient because of the higher capital and operating expenses to separate N_2 production from produced gas streams (N_2 rejection units). Sales gas also might be negatively impacted if an early N_2 breakthrough were experienced, especially in some of the Aeolian formations existing in the area of evaluation. Finally, preliminary economics suggested that other scenarios evaluated would outperform N_2 injection, especially given the high CAPEX and OPEX of N_2 injection projects compared to other possible development scenarios (i.e., gas reinjection and coproduction).

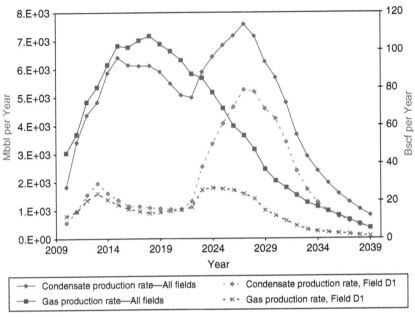

FIGURE 7.9 Yearly condensate and gas production for Field D1 and total for all fields under N_2 injection.

Therefore, at this stage of the evaluation it was possible to identify that, even in the best-case scenario of N_2 injection, the economics did not show enough merit to justify further analysis. This conclusion demonstrates the value of the proposed fast screening methodology in helping the operator to focus on the most valuable options identified.

Case Study, Field E

Field E represented an evaluation of the potential for CO_2-EOR in the United States, CO_2 storage and CO_2 market opportunities in a portfolio of oil reservoirs. The identification of gas fields and saline aquifers as potential CO_2 storage options was also included in this evaluation, but the focus here is mainly on the CO_2-EOR opportunities. Approximately 100 oil reservoirs were identified within 100 square miles from the location of a proposed coal-fired power plant.

It is important to mention that the volume of data to process in this case study was high because of the number of fields identified. However, data coverage varied from field to field, and some fields had insufficient data for a detailed analysis. In these cases, input data were generated from correlations of closer fields producing from the same geologic formations. An important data source used for this purpose was the U.S. DOE Toris database.

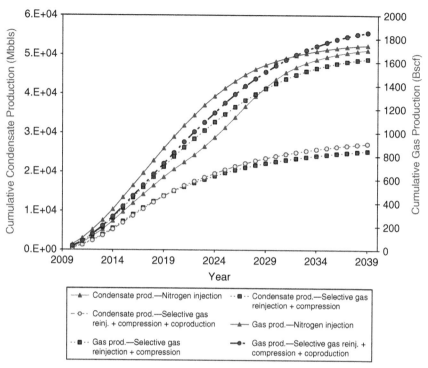

FIGURE 7.10 Cumulative condensate and gas production for Field D for different production scenarios.

To meet project goals and timelines (approximately four months), the analysis focused on identifying potential CO_2 markets for enhanced oil recovery within a small-scale area (50 square miles) and targeting only oil fields with an OOIP greater than 10 million barrels, since smaller fields would not offer large EOR or CO_2 storage potential (Figure 7.11). Taking these constraints into account, it was assumed that if a CO_2 market were justified, it might expand with time once CO_2 distribution pipelines were in place.

A second filter to scope fields for CO_2-EOR was based on current reservoir pressure, oil viscosity, and, most important, current production rates, if available. The purpose of this second fast ranking procedure was to identify fields that showed higher probabilities of CO_2 miscible processes and potentially higher recovery factors (large EOR potential). Based on detailed screening and analytical simulations, a few candidates for CO_2-EOR were identified. Key findings of this screening evaluation indicated that there was an insufficient number of CO_2-EOR opportunities in the vicinity of the proposed coal-fired power plant equipped with CO_2 capture capabilities.

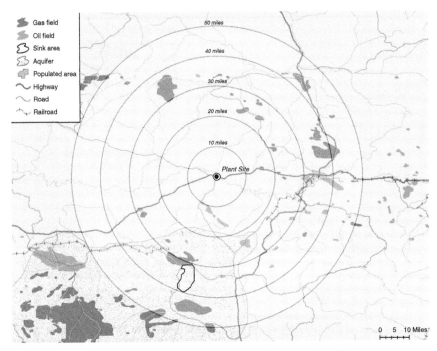

FIGURE 7.11   Plant site location and possible candidates for CO_2-EOR, and storage.

Additionally, by the time any coal-fired plants were built, it would be highly probable that field candidates might not be in operation, and therefore it would be difficult to justify the large capital investments associated with CO_2-EOR projects. In summary, the conclusion of this evaluation strongly suggested that storage opportunities in the area under study should focus on saline aquifers given the volumes expected to be stored and once the proper regulatory framework (e.g., well permitting, CO_2 liability) was in place.

Based on the results of the screening study, a second phase of the project was developed to identify major geologic structures that could store large volumes of CO_2 for geologic time periods. Phase II of the project identified four potential saline aquifers for this purpose. Each of the cases was simulated numerically assuming different scenarios to estimate the number of wells required and plume migration after 300 years for different volumes of CO_2 captured and/or injected. However, the deep saline aquifers identified did not have enough well data to develop an adequate reservoir description.

Therefore, the impact of reservoir heterogeneity on CO_2 injectivity, storability, and plume migration was estimated by randomly generating DP coefficient (homogeneous, base, and heterogeneous cases) distributions using average rock properties and different net pays. The simulation approach contributed to preliminarily ranking the CO_2 sink options identified, providing potential development plans and valuable information—the number of injectors, plume migration versus land ownership, data gathering and monitoring programs, and so forth—that contributed to the decision management strategy to design and prepare the next phase of the project.

Case Study, Field F

The Field F case involved screening and evaluating to support the decision-making process for a property acquisition during the second half of 2006. The property under evaluation was a high-pressure light-oil carbonate reservoir in the United States that had been in operation since the 1970s. In this case study, a detailed screening analysis was carried out to identify the field's mid- and long-term upside potential. However, the most challenging aspect of this study was to develop a reasonable numerical simulation to evaluate multiple production optimization strategies in a short period of time (less than three months).

Neither a full-field geological model nor a numerical model was available at the time of evaluation. However, a large volume of good-quality data (i.e., historical production, core data, multiple reports, etc.) was available to develop the numerical simulation. The latter represented a difficulty given the time frame to decide whether to acquire the field. Therefore, the decision-making process was one of the most challenging steps of this study.

Given the data, budget, and time constraints, different compositional numerical simulation approaches (e.g., history matching and prediction) were discussed. Full-field and large-sector models were discarded. The selection of specific well patterns was also discarded to avoid the selection of an area (i.e., too good or too bad well patterns) not necessarily representative of the entire field. The final decision was to generate a well pattern (element of symmetry) based on the average reservoir properties and the production/injection history of the field and based on detailed reports documented by the operator (seller). This strategy was considered the best option for delivering the results in a timely manner.

The element of symmetry represented approximately 1 percent of the total reservoir volume considering an available detailed lithological and petrophysical interpretation. History match and performance prediction were run using a 9-pseudo-component compositional model. Results were upscaled to yield the same OOIP value for the field, providing expected recovery factor values and injection and production profiles for different scenarios. They were delivered to the client to run the economics. Based on the results and multiple decision risk management sessions the field was purchased. After a year, the field was performing in agreement with the simulation results, and additional upside potentials have been identified from more detailed studies using an integrated reservoir analysis approach.

This chapter presented a few examples that for their emblematic nature illustrate a methodology that the authors conceived as a matter of necessity to deal with complex EOR decision analyses. The strategies discussed should be useful in your situation, but you should adapt the workflow to accommodate the specificities of the particular problems at hand.

8

EOR's Current Status

8.1 INTRODUCTION

This chapter presents a critical review of various field applications for enhanced oil recovery (EOR). The purpose is to provide a context for understanding some of the market drivers for EOR. Paradigmatic cases serve to illustrate how different sides of the financial equation have played a larger role than technical barriers, but the chapter also shows that open literature sources do not necessarily reflect EOR's complete story. Numerous soft issues will be discussed briefly, particularly for current and future uses of EOR technologies for offshore fields.

The focus here is on the drivers of enhanced oil recovery, looking at its evolution in a number of geographical areas. Since this implies a review of bibliographic sources, we are certain that several assertions might not remain valid sometime in the future. However, we believe two facts will remain relevant for quite some time.

First, most of the current world oil production comes from mature fields. Increasing oil recovery from aging resources is a major concern for oil companies and authorities. Second, the rate of replacement of produced reserves by new discoveries has been declining steadily in the last decades, and the trend will be sustained in the foreseeable future. Therefore, increasing the recovery factor in mature fields currently under primary and secondary production will be critical to meeting growing energy demands in coming years.

Enhanced Oil Recovery
DOI: 10.1016/B978-1-85617-855-6.00014-0

Improved oil recovery (IOR) methods encompass enhanced oil recovery (EOR) methods, as well as new drilling and well technologies, intelligent reservoir management and control, advanced reservoir monitoring techniques, and the application of different enhancements of primary and secondary recovery processes, as described earlier. This chapter presents a global overview only of the current status of commercial EOR methods for a spectrum of extra-heavy oil to gas condensate reservoirs. Specifically, the present EOR review is organized by reservoir lithology (sandstone vs. carbonates formations) and offshore versus onshore fields, providing readers with a general idea of actual EOR applications and possible opportunities in future screening and EOR evaluations.

It is well known that EOR projects have been and continue to be strongly influenced by economics and crude oil prices. The initiation of EOR projects depends on the preparedness and willingness of investors to manage EOR risk and economic exposure and the availability of more attractive investment options.

In the United States, chemical and thermal EOR projects declined from the mid-1980s until 2005 (Figure 8.1). However, EOR gas injection projects have exhibited sustained operations since the mid-1980s and have been a growing trend since 2000, especially with the increase in CO_2 projects (Figure 8.2). Indeed, since 2002, EOR gas injection projects

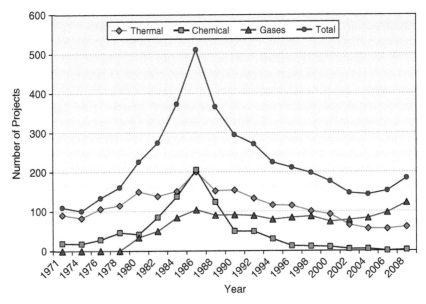

FIGURE 8.1   Trends in U.S. EOR projects. *Source: From* Oil & Gas Journal *EOR Surveys, 1976–2008.*

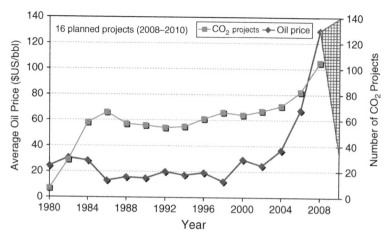

FIGURE 8.2   Evolution of CO_2 projects and oil prices in the United States. *Source: From Oil & Gas Journal EOR Surveys, 1980–2008, and EIA (2009).*

have outnumbered thermal projects for the first time in the last three decades. However, thermal projects have shown a slightly increase since 2004 because of the increase in high-pressure air injection (HPAI) projects in light-oil reservoirs.

Chemical EOR methods still have not captured the interest of oil companies, with only two projects reported in 2008 according to public sources (Aalund, 1988; Bleakley, 1974; Leonard, 1982, 1984, 1986; Matheny, 1980; Moritis, 1990, 1992, 1994, 1996, 1998, 2000, 2002, 2004, 2006, 2008; Noran, 1976, 1978). However, the increase in EOR chemical projects in the United States and abroad has not been documented in the literature for a variety of reasons.

One of the reasons for the increase in EOR gas injection methods in the United States is twofold: (1) availability of vast sources of inexpensive CO_2 from natural sources (US$1–2 per Mscf) and (2) a readily available CO_2 pipeline system, making CO_2-EOR projects economically feasible at oil prices around US$20 per barrel (Manrique et al., 2007; Moritis, 2001). However, it is important to highlight that the CO_2 pipeline system in the United States was built over a 30-year (1975–2005) time span, with favorable oil prices and tax incentives put in place to ensure a secure supply. These are the main drivers for the success of CO_2 operations in a number of basins in the United States, particularly in Texas, as recently reported by Hustad (2009).

Figure 8.2 shows the evolution of CO_2 projects in the United States and average crude oil prices for the last 28 years. Oil prices used for calculating this average are refiner average domestic crude oil acquisition costs, as reported by the Energy Information Administration (2009). For reference purposes, the crude oil price used in Figure 8.2 was arbitrarily selected as that in June of each year.

Although it can be concluded that CO_2-EOR ("from natural sources") is a proven technology that is viable at oil prices greater than \$20 per barrel, this EOR method represents a specific and paradigmatic opportunity in the United States, and it cannot necessarily be concluded that all producing basins in the world with similar fluid and reservoir conditions can be further exploited using this strategy. Therefore, the present review of EOR field experiences incorporates implicitly the importance of technical issues (i.e., reservoir lithology and crude oil viscosity) and nontechnical issues (e.g., onshore vs. offshore, limitations of the existing infrastructure, access to injectants such as CO_2, and the environmental and regulatory framework) in the actual status of international EOR applications.

8.2 EOR BY LITHOLOGY

Reservoir lithology is one of the screening variables for enhanced oil recovery, limiting the applicability of specific EOR methods (Taber et al., 1997a, 1997b). Figure 8.3 shows that most EOR applications have occurred in sandstone reservoirs based on a database from a collection of 1,507 international EOR projects that were consolidated by the authors over the last decade. From Figure 8.3, it should be apparent that EOR thermal and chemical methods are the most frequently used in sandstone reservoirs compared to other lithologies (i.e., carbonates and turbiditic formations).

8.2.1 EOR in Sandstone Formations

EOR methods have been more widely implemented in sandstone formations than in other formation lithologies. In general, from the point of view of experience, sandstone reservoirs show the highest potential for implementation of EOR projects because most of its technologies have been tested at the pilot and commercial scale in this lithology type. Additionally, there are some fields where different EOR technologies have been evaluated successfully at the pilot scale, demonstrating the technical applicability of different EOR methods in the same field. Buracica and Carmópolis (Brazil) and Karazhanbas (Kazakhstan) are good examples of fields where different EOR methods were tested at the pilot scale in sandstone formations; they are described in the following list:

- *Buracica* is an onshore light-oil (34–37°API) reservoir with reported air injection (1978–1980), immiscible CO_2 injection (1991), and polymer flooding (1997) pilot projects. The operator reported the expansion of immiscible CO_2 injection (de Melo et al., 2002, 2005; Lino, 2005; Schecaira et al., 2002).
- *Carmópolis* is the largest onshore field in Brazil that contains heavy oil (22°API), with reported in situ combustion (1978–1989), polymer flooding (1969–1972 and 1997), steam injection (1978), and microbial EOR or MEOR (2002) pilot projects. The field has been developed mainly by waterflooding (da Silva et al., 2007; de Melo et al., 2005; de Souza et al., 2005; Mezzomo et al., 2001; Pinto et al., 2006).
- *Karazhanbas* is an onshore heavy-oil (19°API) reservoir with documented polymer flooding (Chakabaev et al., 1978), steam injection (Antoniadi et al., 1993; Mamedov and Bokserman, 1992), and in situ combustion and in situ combustion plus foam injection as a conformance strategy (Antoniadi et al., 1993; Zhdanov et al., 1996). Karazhanbas field was developed by waterflooding, cold heavy-oil production with sand (CHOPS) (Collins et al., 2008), and steam injection.

As shown in Figure 8.3, thermal and chemical methods consist mostly of EOR applied in sandstone formations compared to EOR gas injection. The next section provides an overview of different EOR methods implemented in sandstone formations.

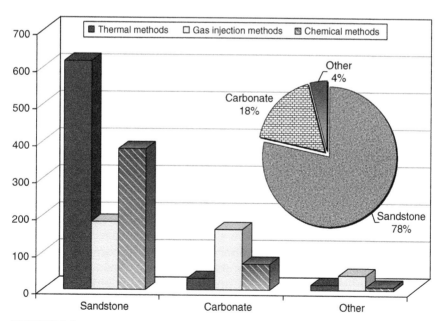

FIGURE 8.3   EOR methods by lithology for a total of 1,507 projects.

Thermal Methods

Cyclic steam injection (i.e., steam stimulation, or huff and puff), steam flooding, and, most recently, steam-assisted gravity drainage (SAGD) have been the most frequently used recovery methods for heavy- and extra-heavy-oil production in sandstone reservoirs during the last decades. Thermal EOR projects are mostly found in Canada; the former Soviet Union (FSU); the United States; Venezuela; Brazil; and, to a lesser extent, China.

Steam injection began approximately five decades ago. The Mene Grande and Tia Juana fields in Venezuela (De Haan and Van Lookeren, 1969; Ernandez, 2009) and Yorba Linda and Kern River fields in California (Hanzlik, 2003) are good examples of steam injection projects within the last four decades. Some more recent examples reported in the literature are the steam floods in the Crude E field in Trinidad (Ramlal, 2004), Schoonebeek oil field in the Netherlands (Jelgersma, 2007), and Alto do Rodrigues in Brazil (Lacerda et al., 2008). Although optimization of steam injection processes has been attempted through the use of solvents (Rivero and Mamora, 2007), gases (Bagci and Gumrah, 2004), chemical additives (Ovalles et al., 2001), and foams (Mendez et al., 1992), among others; only a few of these proposed methods have been tested in the field (Mbaba and Caballero, 1983; Mendez et al., 1992; Zhdanov et al., 1996).

One example is the liquid addition to steam for enhancing recovery (LASER) process, which involves the injection of C_5+ liquids as a steam additive in cyclic steam injection processes. Although the LASER process was tested at a pilot scale in Cold Lake (Leaute, 2002), it has not been expanded at a commercial scale. Steam injection has also been tested in medium- and light-oil reservoirs, where crude oil distillation and thermal expansion are the main recovery mechanisms (Perez-Perez et al., 2001). However, steam injection in medium- and light-oil reservoirs has not contributed significantly to EOR production worldwide.

Steam-assisted gravity drainage represents another important EOR thermal method aimed at increasing oil production in oil sands. Because of the applicability of SAGD in unconsolidated reservoirs with high vertical permeability (Manrique and Pereira, 2007), this EOR method has received attention in countries with heavy- and extra-heavy oil resources, especially Canada and Venezuela, that own vast oil sands resources. However, and despite SAGD pilot tests reported in China (Li-qiang et al., 2006), the United States (Grills et al., 2002), and Venezuela (Mendoza et al., 1999), commercial applications of SAGD have been reported in Canada only, more specifically in the McMurray Formation, Athabasca (e.g., Hanginstone, Foster Creek, Christina Lake, and Firebag).

Figure 8.4 shows reservoir depths, average horizontal permeability, and the formation of several SAGD (pilot- and commercial- scale) projects

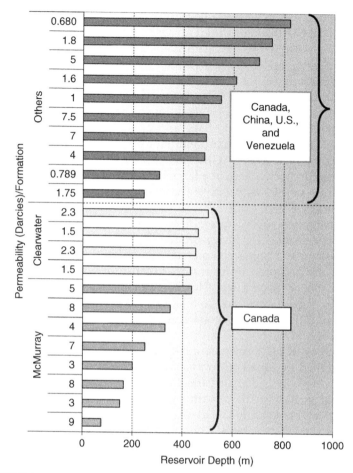

FIGURE 8.4 Depth, average permeability, and formation of SAGD field projects.

that are well documented in the literature. Among these projects, only those developed in the McMurray Formation (bottom group of bars in figure) operate commercial SAGD projects. SAGD projects tested in the Clearwater Formation in Cold Lake, Canada (light bars in figure) have proved to be uneconomical (Scott, 2002).

Commercial SAGD projects in the McMurray Formation validate the importance of the geology and reservoir characteristics in this enhanced oil recovery method. This finding has been reported by Rottenfusser and Ranger (2004), Putnam and Christensen (2004), and Jimenez (2008), among others. Therefore, the present level of understanding of the SAGD process and field experience strongly suggest that this technology will continue to expand, depending of course on oil price projections, mainly in Athabasca's McMurray Formation.

Alternatives to SAGD have been proposed. They include modified versions of SAGD through different well configurations or numbers of wells (e.g., cross or X-SAGD, fast SAGD, and single-well SAGD or SW-SAGD) and using additives (e.g., ES-SAGD) in steam, as reported by a number of authors (Elliot and Kovscek, 2001; Govind et al., 2008; Shin and Polikar, 2005; Stalder, 2008).

However, all of the proposed methods are at the early stage of evaluation and are not expected to have an impact on oil production in the near future. This is an important reminder for readers who may be unfamiliar with the history of SAGD. This process took roughly 30 years to become a mature technology in a large basin in Canada. Perhaps the fate of alternatives will be as positive and their full implementation as fast, but only time will tell.

In situ combustion (ISC) projects have been the second most important recovery method for heavy crude oils in the past decades. Despite its long history and some commercial successes, this EOR process has not been fully accepted by operators because of an excessive number of inconclusive or failed pilots. However, an important number of failed projects can also be attributed to a lack of understanding of the process and to applications in reservoirs not necessarily appropriate for this EOR process.

Although a few ISC projects are ongoing in heavy oil reservoirs—such as Battrum field in Canada (Moritis, 2008); Suplacu de Barcu in Romania (Machedon et al., 1995; Panait-Paticaf et al., 2006); Balol, Bechraji, Lanwa, and Santhal in India (Chattopadhyay et al., 2004; Doraiah et al., 2007; Moritis, 2008; Roychaudhury et al., 1997; Sharma et al., 2003); and Bellevue in the United States (Long and Nuar, 1982; Moritis, 2008)—air injection in light-oil reservoirs (referred to as high-pressure air injection, or HPAI) has gained greater attention during the last decade.

The successful application of air injection projects in light-oil reservoirs like West Hackberry in the United States demonstrates that this recovery process is a viable EOR strategy for high-dipping-angle reservoirs combined with double displacement (DDP) strategies (Gillham et al., 1997, 1998). Since 2000 the number of ISC projects has been steady, as reported by Moritis (2008), with ten projects in sandstone formations. At the same time, the number of HPAI projects in U.S. light-oil reservoirs has shown important growth during the same period (Figure 8.5). However, all of these HPAI projects have been implemented in carbonate formations, which will be discussed later in the chapter.

In addition to the increasing ISC and HPAI trends reported by Moritis (2008) during this decade, Duiveman and colleagues (2005) and Hongmin and colleagues (2008) documented air injection projects in Handil field in Indonesia (2001), and Hu 12 Block, Zhong Yuan field in China, respectively. Although the Handil field HPAI pilot (0.5–1 cp oil) reported

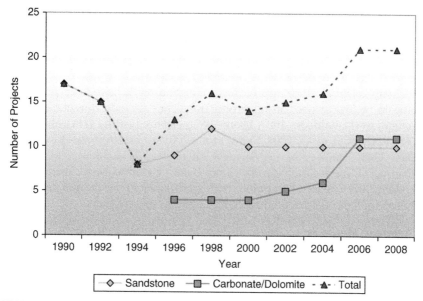

FIGURE 8.5   Trends in ISC and HPAI. *Source: From Moritis, 2008.*

injectivity problems due to lack of reservoir communication in the pilot area, the results were reported as encouraging (Duiveman et al., 2005).

On the other hand, the report about the pilot in Zhong Yuan field (3.9 cp oil), which consisted of the use of foam-assisted water-alternating air injection, also documented encouraging results (Hongmin et al., 2008). Other examples illustrate the increased interest in air injection. These include the planned ISC in Rio Preto, west onshore Brazil, reported by Moritis (2008), and studies done by Hughes and Sarma (2006), Sarma and Das (2009), Teramoto and colleagues (2005), and Onishi and colleagues (2007) evaluated the technical feasibility and potential of HPAI in Australia and Japan, respectively. Based on recent trends, we suggest that air injection, especially in light-oil reservoirs (HPAI), will continue to grow in the next decade.

Alternatives to ISC, such as toe-to-heel air injection, or THAI (Greaves et al., 2005; Xia et al., 2002), and CAPRI (Xia et al., 2002) have been proposed. CAPRI is the catalytic version of THAI (catalytic THAI). Both processes are at early stages of evaluation through the Whitesands project partnership as reported by Petrobank Energy and Resources Ltd. (2009). Results for this project might be available sometime during late 2010 to confirm its EOR potential in oil sands and its potential applicability to other types of heavy-oil reservoirs.

THAI is not expected to have an impact on EOR production in the near future. Additionally, and as usual, crude oil price volatility will continue to play a key role, justifying further THAI pilot tests. With regard to the CAPRI processes, current competing surface-upgrading technologies (e.g., SINCOR in Venezuela and Long Lake in Canada) may not justify the use of a hydrotreating catalyst or costly hydrogen donors without operational experiences with THAI.

Finally, several other approaches to thermal EOR have been proposed with no or low impact on oil production. They include downhole steam generation (Donaldson, 1997; Eson, 1982), electric heating (Hascakir et al., 2008; Rodriguez et al., 2008; Sierra et al., 2001) or electromagnetic heating (Das, 2008; Islam et al., 1991), and microwave (Hascakir et al., 2008). However, these technologies have not been proved to be technically and economically feasible compared with traditional EOR thermal methods. Therefore, they will not be further discussed in this chapter.

Chemical Methods

EOR chemical methods had their best times in the 1980s, most of them in sandstone reservoirs (Needham and Doe, 1987). The total number of active projects peaked in 1986, with polymer flooding as the most important among them (refer to Figure 8.1). However, since the 1990s, oil production from EOR chemical methods has been negligible around the world except for China (Chang et al., 2006; Delamaide et al., 1994; Han et al., 1999; Li et al., 2009; Wang et al., 2002, 2009a, 2009b; Xiaoqin et al., 2009). Nevertheless, chemical flooding has been shown to be sensitive to the volatility of oil markets despite recent advances (e.g., low-surfactant concentrations) and lower costs of chemical additives.

Polymer flooding should be considered a mature technology, and it is still the most important EOR chemical method in sandstone reservoirs, based on a review of full-field case histories. It is important to point out that this section does not consider near-wellbore treatments (i.e., gels and polymer-gels) as EOR processes because we view this type of treatments as beyond our scope. Based on the EOR survey presented by Moritis (2008), there are ongoing pilots or large-scale polymer floods in Argentina (El Tordillo field), Canada (Pelican Lake), China with approximately 20 projects (e.g., Daqing, Gudao, Gudong, and Karamay fields, among others), India (Jhalora field), and the United States (North Burbank).

It is important to mention that a commercial polymer flood was developed in North Burbank during the 1980s (Moffitt and Mitchell, 1983), demonstrating that this EOR method may still have the potential to increase oil recovery in mature basins (e.g., mature floods with movable and/or by-passed oil). North Burbank reinitiated polymer flooding on a 19-well pattern in December 2007 (Chaparral Energy Inc., 2009). Other countries with reported polymer flooding projects include the Brazilian Carmópolis, Buracica, and Canto do Amaro fields (Shecaira et al., 2002).

India also reported a polymer flood in Sanand field (Pratap et al., 1997; Tiwari et al., 2008). Oman documented a polymer flood pilot that was developed in Marmul field (Koning et al., 1988), and almost 20 years later, a large-scale application is underway (Moritis, 2008).

Additionally, Austria's Pirawarth field (Poellitzer et al., 2009), Argentina (El Tordillo field), Brazil (Voador offshore field), Canada (Horsefly Lake field), and Germany (Bochstedt field) announced plans to implement polymer flood projects (Moritis, 2008). Listed ongoing and planned polymer floods provide a representative sample of field experiences that validate the EOR potential of this recovery process. Small sandy reservoirs, such as those found in the United States in Wyoming (Alvarado et al., 2008), have been flooded with either straight (polymeric blends) or more complex chemical designs.

Colloidal dispersion gels (CDGs) and BrightWater® also represent novel polymer-based technologies that are currently under evaluation at field scales. Although these technologies are quite different from the technical standpoint, both are meant to improve volumetric sweep efficiency in mature waterfloods, especially in reservoirs with high-permeability contrasts and the presence of thief zones. Documented CDG projects include Daqing field in China (Chang, Sui et al., 2006; Chang, Zhang et al., 2006; Mack, 2005) and El Tordillo (Diaz et al., 2008) and Loma Alta Sur fields in Argentina (Muruaga et al., 2008).

Regarding BrightWater (Frampton et al., 2004), at the present time, Milne Point in Alaska is the only field application discussed or documented in the public domain (Ohms et al., 2009; Puckett, 2009). However, it is expected that the number of CDGs and BrightWater field applications will increase in the near future, based on recent field and laboratory studies now underway, opening a new window of opportunities for EOR chemical methods (Manrique, 2009; Spildo et al., 2009).

Although polymer flooding has been the most applied EOR chemical method in sandstone reservoirs (Manning et al., 1983), injections of alkali, surfactant, alkali-polymer (AP), surfactant-polymer (SP), and alkali-surfactant-polymer (ASP) have been tested in a limited number of fields (refer to Figure 8.3). Micellar polymer flooding had been the second most frequently used EOR chemical method in light and medium crude oil reservoirs until the early 1990s (Lowry et al., 1986). Although this recovery method had been considered a promising EOR process since the 1970s, high concentrations and costs of surfactants and cosurfactants, combined with low oil prices during mid-1980s, limited its use. The development of ASP technology since the mid-1980s and the development of surfactant chemistry have brought renewed attention to chemical floods in recent years, especially to boost oil production in mature and waterflooded fields.

Several EOR chemical methods, other than polymer floods, have been widely documented in the literature during the last two decades. However, at present, Daqing field represents one of the largest ASP floods, if not the largest, implemented as of today. ASP flooding has been studied and tested in Daqing for more than 15 years, through use of several pilots of different scales (Chang et al., 2006; Demin et al., 1999; Hongfu et al., 2003; Li et al., 2009; Pu and Xu, 2009).

The Gudong (Qu et al., 1998), Karamay (Gu et al., 1998; Qiao et al., 2000), Liahoe, and Shengli (Chang et al., 2006) fields are all examples of Chinese ASP projects that have been documented in the literature. Additional EOR chemical floodings reported during the last decade include the following:

- ASP flooding in Viraj field in India (Pratap and Gauma, 2004) and in the U.S. fields: West Kiehl (Meyers et al., 1992), Sho-Vel-Tum (French, 1999), Cambridge Minnelusa (Vargo et al., 2000), and Tanner (Pitts et al., 2006).
- AP flooding in Xing Long Tai oil field (Zhang et al., 1999) in China and David Pool field in Canada (Pitts et al., 2004).

Based on the EOR survey presented by Moritis (2008), ongoing ASP pilots exist in Delaware Childers field (Oklahoma) and Lawrence field (Illinois). ASP floods are planned in Lawrence field, and Nowata field (Oklahoma), along with an SP in Midland Farm Unit, Texas (Moritis, 2008) and in Minas field in Indonesia (Bou-Mikael et al., 2000). However, the number of ASP and SP floods is much higher than that reported in the literature and in the EOR survey presented by Moritis (2008) because operators do not necessarily respond to this survey.

The authors are aware of ongoing projects in the United States and Canada that have not yet been published in the literature. Additionally, several projects are ongoing in Argentina, Canada, India, and the United States under reservoir and lab evaluations, with pilot projects scheduled between 2010 and 2011. Therefore, and despite the volatility of oil prices, it is fair to conclude that operators are showing a growing interest in EOR chemical flooding. This trend is also noticed in the increase in screening studies to evaluate or reestimate the EOR potential of chemical flooding in different basins (Alvarado et al., 2008; Costa et al., 2008; Fletcher and Morrison, 2008; Pandey et al., 2009).

Gas Methods

EOR gas flooding has been the most widely used recovery method in light, condensate, and volatile oil reservoirs. Although nitrogen (N_2) injection has been proposed to increase oil recovery under miscible conditions, favoring the vaporization of light fractions of light oils and condensates, today only a few N_2 floods are ongoing in sandstone reservoirs. Immiscible N_2 floods are reported in Hawkins field (Texas) and Elk Hills field (California), based on the Moritis EOR survey (2008). No new N_2 floods in sandstone reservoirs have been found in the open literature during the last few years, and we do not foresee an increment in the number of projects implementing this EOR gas flooding method.

Similarly to N_2 injection, hydrocarbon gas injection projects conducted in onshore sandstone reservoirs have made a relatively marginal contribution, in terms of total oil recovered, in Canada and the United States. It is important to mention that, in this book, we refer to water-alternating-gas (WAG) injection schemes, enriched gases, or solvents and their combinations as EOR gas methods using hydrocarbon gases. Therefore, hydrocarbon gas injections as pressure maintenance or double-displacement strategies are not considered EOR methods for purposes of this review.

Most of the immiscible and miscible EOR hydrocarbon gas floods in the United States are on the North Slope of Alaska (Moritis, 2008; Panda et al., 2009; Rathman et al., 2006; Redman, 2002; Shi et al., 2008), while in Canada, a miscible gas flood is reported in Brassey field (Moritis, 2008). The state of hydrocarbon gas injection projects, discussed later in this chapter, is different in offshore sandstone reservoirs (Christensen et al., 2001). In general, if there is no other way to monetize natural gas, then a more practical use of it would be in pressure maintenance projects or in WAG processes.

However, and if they are available, the substitution of hydrocarbon gases by nonhydrocarbon gases (N_2, CO_2, acid gas, air) for oil recovery will make more natural gas available for domestic use or export while still maintaining reservoir pressure and increasing oil recoveries. Despite the current volatility of natural gas prices, the continuous increase in energy demand will likely affect the viability of new large-scale hydrocarbon gas projects.

On the other hand, CO_2 flooding has been the most widely used EOR recovery method for medium- and light-oil production in sandstone reservoirs during the last few decades, especially in the United States, because of the availability of cheap and readily available CO_2 from natural sources. Earlier in this chapter, Figure 8.2 clearly showed an increasing trend in CO_2 field projects in the United States during the last decade in both sandstone and carbonate reservoirs. The number of CO_2 floods is expected to continue growing in U.S. sandstone reservoirs.

Some examples of planned CO_2-EOR projects in the United States include Cranfield, Heidelberg West, and Lazy Creek fields in Mississippi and Sussex field in Wyoming (Moritis, 2008). The number of CO_2 floods in Wyoming sandstone reservoirs is also expected to increase based on a recent evaluation presented by Wo and colleagues (2009). Additionally, Holtz (2008) reported only recently on an overview of sandstone gulf coast and Louisiana CO_2-EOR projects to estimate EOR reserve growth potential in the area, including sandstone reservoirs in the Gulf of Mexico. CO_2-EOR in the United States has shown a vast potential to increase oil recovery and has been widely documented in the literature, so the following paragraphs briefly address activities reported only outside the United States.

Some examples of CO_2-EOR field projects in sandstone formations presented at various conferences and/or documented in the literature follow:

- Brazil reported CO_2 floods in Buracica and Rio Pojuca fields (Dino et al., 2007; Moritis, 2008) and announced a CO_2 flood in Miranga field from anthropogenic sources as an EOR and carbon storage strategy (Dino et al., 2007; Guedes, 2008).
- Canada actually reported CO_2 floods in Joffre and Pembina fields (Moritis, 2008; Stephenson et al., 1993). Canadian operators and government institutions have been very active in evaluating CO_2-EOR potential during the last decade (Bachu et al., 2000; PTAC, 2003). Recently, PTAC (Petroleum Technology Alliance of Canada) estimated an upside potential of CO_2-EOR in Alberta of 3.6 billion barrels over the next two decades at oil prices of $45/bbl (Byfield, 2009).
- Croatia reported CO_2 pilot injection at Ivanić field, injecting CO_2 transported by trucks. Pilot results (2001–2006) contributed to defining larger applications of CO_2-EOR in the country considering the use of CO_2 from anthropogenic sources (Domitrović et al., 2004; Novosel, 2005, 2009).
- Hungary reported more than four decades of experience in CO_2-EOR floods. CO_2 projects at Budafa and Lovvaszi fields are two cases that are well documented in the literature (Doleschall et al., 1992). Szank oil field represents a more recent CO_2-EOR flood in Hungary using CO_2 from a sweetening plant (Remenyi et al., 1995).
- Trinidad has more than three decades of experience operating CO_2-EOR projects using CO_2 from an ammonia plant near the fields (Mohammed-Singh and Singhal, 2004). A Moritis EOR survey (2008) reports up to nine active immiscible CO_2 floods operating since the mid-1970s.

As can be seen, CO_2-EOR has become one of the preferred EOR processes globally, using CO_2 from both natural and industrial sources. Mexico (Bauer, 2006; Muro et al., 2007) and the United States (Kulkarni et al., 2008; Wo et al., 2009) are just two examples of countries evaluating CO_2 sources and EOR potential in mature fields and mature CO_2 floods (Senocak et al., 2008). However, this will be further discussed in the next section.

8.2.2 EOR in Carbonate Formations

It is well known that a considerable portion of the world's hydrocarbon endowment is in carbonate reservoirs. Carbonate reservoirs usually exhibit low porosity and may be fractured. These two characteristics, along with oil-to-mixed wet rock properties, usually result in lowered hydrocarbon recovery rates. When EOR strategies are pursued, the injected fluids likely flow through the fracture network and bypass the oil in the rock matrix. The high permeability in the fracture network and the low equivalent porous volume result in an early breakthrough of the injected fluids.

A large number of EOR field projects in carbonate reservoirs have been referenced in the literature during the last few decades. Although these projects demonstrate the technical feasibility of various EOR methods in carbonate reservoirs, gas injection (continuous or in WAG mode) is still the most common implemented in this type of lithology (see Figure 8.3). Polymer flooding is the only proven EOR chemical method in carbonate formations, while EOR thermal methods have made a relatively small contribution to the world's oil production from carbonate reservoirs. However, the number of high-pressure air injection (HPAI) projects has increased in recent years, especially in light-oil carbonate reservoirs in the United States (Manrique, 2009).

As opposed to sandstone reservoirs, in some fields different EOR technologies have been evaluated successfully at the pilot scale, demonstrating the technical applicability of different EOR methods in carbonate formations. The Yates field (Texas) represents a good example of a carbonate formation where various EOR processes have been tested successfully at different scales (from pilot- to large-scale applications).

The following are some of the EOR processes evaluated in the Yates field that have been documented in the literature:

- Nitrogen (N_2) injection began in the mid-1980s as a reservoir pressure maintenance strategy (Button and Peterson, 2004; Clancy et al., 1985; Levine et al., 2002).
- Steam flooding (pilot) was initiated late in 1998 as a potential strategy to improve vertical gravity drainage (Button and Peterson, 2004; Dehghani and Ehrlich, 1998; Snell and Close, 1999; Snell et al., 2000).
- A dilute surfactant well-stimulation pilot test was reported in the early 1990s as a strategy to increase oil recovery by IFT reduction, gravity segregation of oil, and wettability alteration, among other mechanisms (Button and Peterson, 2004; Campanella et al., 2000; Chen et al., 2000, 2001; Manrique et al., 2007; Yang and Wadleigh, 2000).

- In March 2004, Yates field started replacing N_2 injection with CO_2 injection as a pressure maintenance strategy and to enhance gravity drainage (Button and Peterson, 2004).

Manrique and colleagues (2007) presented a comprehensive review of EOR field experiences in U.S. carbonate reservoirs. Although this review was specific to carbonate formations, it can be considered representative of the estimation of technical feasibility and the potential for EOR processes in this type of reservoir based on valuable field experiences documented in the literature.

Alvarez and colleagues (2008) published a literature review of field experiences specifically in heavy-oil carbonate reservoirs, including several pilot tests carried out in the Grosmont Formation in Canada during the 1970s and 1980s (i.e., Chipewyan River, Buffalo Creek, McLean, Orchid, Saleski, and Algar). The following sections provide a general overview of different EOR methods implemented in carbonate formations around the world.

Thermal Methods

Thermal EOR projects have not been popular in carbonate formations. Neither cyclic nor continuous steam injections have been widely used in carbonate reservoirs. The Garland field in Wyoming (Dehghani and Ehrlich, 1998) and the Yates field in Texas (Button and Peterson, 2004) represent two of the few steam drive projects in carbonate formations documented in the United States (Manrique et al., 2007). The following are some of the steam injection projects documented in carbonate formations outside of Canada and the United States:

- Steam drive pilot at Lacq Supérieur field in France (Perez-Perez et al., 2001; Sahuquet and Ferrier, 1982; Sahuquet et al., 1990)
- Steam flood pilot at Ikiztepe field, a heavy-oil fracture reservoir in Turkey (Nakamura et al., 1995)
- Cyclic steam pilot in Cao-32 field, a fracture limestone heavy crude oil reservoir in China (Zhou et al., 1998)
- Steam flood pilot, and recently announced full-field implementation, in Qarn Alam field in Oman (Macaulay et al., 1995; Moritis, 2008; Penney et al., 2005, 2007). Oman also announced a steam injection project in the limestone Fahud field, among other steam pilots (Moritis, 2008).
- Cyclic steam injection pilot in Issaran heavy-oil field in Egypt (Buza, 2008; Waheed et al., 2001)
- Steam flood pilot at giant Wafra field in Kuwait and Saudi Arabia (Barge et al., 2009; Buza, 2008).

As can be seen, steam injection in carbonates has been mostly tested at the small scale, and only Qarn Alam field in Oman is announcing full-field steam flooding operations. Therefore, steam injection at Qarn Alam may contribute to defining the future of steam injection in carbonate formations.

SAGD is another technology that has been proposed for carbonate reservoirs (Sedaee and Rashidi, 2006; Shafiei et al., 2007), and a very limited number of studies are considering this recovery process for fractured carbonate reservoirs. However, we believe the fractured and vuggy nature of carbonate formations can cause uneven sweep along SAGD well pairs. This may lead to the irregular development of steam chambers, causing the early breakthrough of steam into the horizontal producer and resulting in a low recovery factor—and therefore an uneconomical project.

On the other hand, air injection projects in carbonate formations have shown a steady increase since 2000, especially HPAI projects in U.S. light-oil reservoirs (refer to Figure 8.5). To date, 11 active HPAI projects are running in light-oil (>30°API) carbonate reservoirs in Montana and South and West Buffalo and Medicine Pole Hill in North and South Dakota (Moritis, 2008). These are all good examples of combustion projects in light crude oil dolomitic formations (Manrique et al., 2007).

The Buffalo field (North Dakota) started air injection approximately three decades ago, and projects are still in operation (Gutiérrez et al., 2008; Kumar et al., 2008). The success and expansion of Buffalo and of Medicine Pole Hill (South Dakota) have contributed to the increase in HPAI projects in the area. Although all of the HPAI projects reported by Moritis (2008) in Montana and North and South Dakota have been developed in the same low-permeability dolomitic formation (Red River A, B, and/or C), air injection has proved to have high potential for improving oil recovery and revitalizing both mature and waterflooded carbonate reservoirs (Gutiérrez et al., 2008; Moore et al., 2002). This will be especially true in fields located in remote locations with no access to CO_2 sources.

There is no doubt that risk perception with air injection processes is still part of our industry. However, current HPAI projects in U.S. carbonate reservoirs are demonstrating that risks can be mitigated and that projects can be economically attractive. Mexico is one country that is evaluating air injection processes in naturally fractured carbonates, given that most of its production and reserves are coming from this reservoir type. Mexico announced a potential HPAI project in the Cárdenas field, an onshore light-oil (40°API) fractured carbonate reservoir located in the southern region of the Chiapas-Tabasco basin (Rodríguez and Christopher, 2004). Therefore, production results from recent U.S. air injection projects (Williston Basin) and potentially the pilot project in

Cárdenas field (Mexico) are likely to dictate the future of this recovery method in carbonate reservoirs in the United States and abroad.

Chemical Methods

Polymer flooding is the only proven chemical EOR technology, mostly at the early stages of waterflooding, in carbonate reservoirs (see Figure 8.3). However, carbonate reservoirs have made a relatively small contribution to polymer flooding in terms of total oil recovered in the United States (Manrique et al., 2007). With today's technology, alkali-polymer (AP) and alkali-surfactant-polymer (ASP) floods are applicable to sandstone reservoirs only. However, surfactant-polymer (SP) seems to be a feasible recovery process in both carbonate and sandstone reservoirs.

To date, no chemical flooding other than polymer flooding in carbonate reservoirs has been reported in the literature reviewed. However, ASP has been tested in carbonate formations in the lab for Arab-D (Al-Hashim et al., 1996), Upper Edward (Manrique et al., 2007), and Pietra Leccese outcrop (Bortolotti et al., 2009) core samples. An alkali-surfactant single-well test was reported in the Mauddud carbonate reservoir in Bahrain as part of ASP feasibility studies in oil-wet limestone reservoirs (Zubari and Sivakumar, 2003).

Surfactant injection was the only chemical method used recently as a well stimulation and wettability modifier in carbonate reservoirs. In fractured reservoirs, spontaneous water imbibition can occur from fractures into the rock matrix. Subsequently, this mechanism leads to oil drainage from the matrix toward the fracture network, making surfactants favorable for improving oil recovery in oil-wet carbonate reservoirs by changing rock wettability (to mixed/water-wet) and promoting the imbibition process.

Cottonwood Creek and Yates fields are two examples of surfactant stimulation wells that have been documented in the literature. Recent studies in Norway have demonstrated the positive effect of the injection of seawater into carbonate chalk reservoirs, where the sulfate-containing seawater contributes to the development of strong water wettability in the chalk matrix rock over the years of waterflooding (Manrique et al., 2007).

Given the vast quantity of the world's oil reserves that exist in carbonate reservoirs, chemically assisted methods (i.e., spontaneous imbibition, wettability modifiers, and ITF reductions) based on surfactant injections represent an active research area as a strategy for improving oil recovery in carbonate formations (Adibhatia and Mohanty, 2007; Delshad et al., 2009; Haugen et al., 2008; Mohanty, 2006; Najafabadi et al., 2008; Tabary et al., 2009; Webb et al., 2005). However, as of mid-2010, no large field application has been documented in the literature.

Based on the current status of the technology, EOR chemical methods are not expected to make an important contribution to oil production from carbonate reservoirs during the next decade or two. However, chemically based gas and water shutoff strategies (i.e., gels and foams) will continue to contribute to optimizing water, gas, or WAG projects in carbonate reservoirs in the near future (Al-Dhafeeri et al., 2008; Al-Taq et al., 2008; Hirasaki et al., 2006; Kumar et al., 2007; Portwood, 2005; Sengupta et al., 2001; Smith et al., 2006).

Gas Methods

EOR gas flooding has been the most widely used recovery method in light, condensate, and volatile oil carbonate reservoirs. Figure 8.3 earlier in the chapter clearly showed that gas injection has been the enhanced oil recovery method most frequently applied in carbonate formations compared to EOR chemical and thermal methods.

Nitrogen (N_2) flooding has been an effective recovery process for deep, high-pressure, and light-oil reservoirs. Generally for these types of reservoirs, nitrogen flooding can reach miscible conditions. However, immiscible N_2 injection has also been used for pressure maintenance and cycling of condensate reservoirs, and as a gas drive for miscible slugs (Manrique et al., 2007). Although over the last four decades, several N_2 injection projects have been reported in carbonate reservoirs in the United States, Moritis (2008) reports only one miscible WAG: N_2 in Jay LEC (Lawrence et al., 2002). However, more recently, the operator of that field announced a temporary production suspension (SEC, 2009).

Outside the United States, Cantarell is the only representative N_2 injection project ongoing in an offshore carbonate field that is well documented in the literature (Sánchez et al., 2005). No new N_2 floods in carbonate reservoirs have been reported during the last few years, and we do not foresee an increase in the number of projects implementing this EOR gas-flooding method, except perhaps for small units. High capital costs (e.g., air separation units) and operational costs (e.g., N_2 rejection units, if required) associated with N_2 injection have reduced interest in this recovery process in recent years.

Additionally, with the recent successes and field expansions reported in Montana, North Dakota, and South Dakota, HPAI (high-pressure air injection) has surged as a potential option with inherent lower capital and operational costs than those for miscible N_2 floods. However, N_2 injection still represents an option that can be justified for high-pressure and high-temperature (HP/HT) light-oil reservoirs if no access to any other gas sources is possible (Mungan, 2000).

Similar to N_2 injection, hydrocarbon gas injection in onshore carbonate reservoirs has made a relatively marginal contribution in terms of total oil recovered in Canada and the United States (Manrique, et al., 2007; Moritis, 2008). It is important to mention that in this book we refer to EOR gas methods using hydrocarbon gases water-alternating-gas (WAG) injection schemes, enriched gases or solvents, and their combinations (hydrocarbon miscible flooding) (Stalkup, 1983). Therefore, immiscible hydrocarbon gas injections as pressure maintenance or double-displacement strategies are not considered EOR methods for our purposes.

Some examples of documented hydrocarbon miscible flooding (continuous injection or WAG mode) ongoing, or under evaluation in carbonate formations, are reported in Canada, the Middle East, and various offshore locations (Al-Bahar et al., 2004; Christensen et al., 2001; Edwards, 2002; El Mahdi et al., 2007; Gomes et al., 2002; Mijnssen et al., 2003; Moritis, 2008; Schneider and Shi, 2005). If there is no other way to monetize natural gas, then a more practical use for it would be in pressure maintenance projects or in WAG processes as new business opportunities become available. This development strategy will help preserve reservoir energy, maximizing oil recovery with an upside potential of natural gas through depressurization strategies late in the reservoir's production life.

CO_2-EOR has been successfully implemented in both mature and waterflooded carbonate reservoirs (Manrique et al., 2007; Moritis, 2008). CO_2 flooding from natural sources has been the most important EOR process in the United States, particularly in carbonate reservoirs of the Permian Basin. Moritis (2008) reported 105 active CO_2 floods, 63 of them in carbonate formations mainly in the Texas Permian Basin. As we saw before, the popularity of CO_2 projects is closely related to the abundant availability of natural sources of CO_2 and associated CO_2 transporting pipelines that are generally located close to the oil fields (Hustad, 2009; Manrique et al., 2007).

CO_2-EOR in U.S. carbonate reservoirs is expected to continue to grow (refer to Figure 8.2) based on natural sources of CO_2. If CO_2 flooding is to increase, nonnatural sources will need to be incorporated at competitive costs. In simple terms, if CO_2 is available, it will remain the most sound recovery choice for carbonate reservoirs unless more viable EOR strategies are developed. Canada (i.e., Enchant Midale, Judy Creek, Swan Hills, and Weyburn) and Turkey (e.g., Bati Raman) also report well-documented CO_2-EOR projects in carbonate formations (Asghari et al., 2007; Karaoguz et al., 2007; Louie, 2009; Moritis, 2008; Sahin et al., 2008).

Climate change has become an issue of intense debate during the last decade. Despite heated disagreement as to whether global warming is linked to population growth and industrial development, the international scientific community is proactively trying to secure resources to meet future energy demands while simultaneously restricting CO_2 and other greenhouse gas emissions generated by energy production (Hamilton, 2009). This topic has been overwhelmingly documented in the literature. Thus, this review will not try to provide an extensive list of references related to it. What the authors are trying to highlight is that CO_2-EOR has become an attractive CO_2 storage alternative among the options currently available. However, it is important to point out that CO_2 storage capacity in oil and gas reservoirs is limited (Manrique, 2009; Manrique and Araya, 2008).

The fundamental reason why CO_2 sequestration combined with EOR has become the preferred emission reduction strategy is that high hydrocarbon price scenarios provide the necessary financial incentive for increasing oil and gas reserves through EOR methods and also generate the capital needed to fund such projects until a suitable regulatory framework is in place (Algharaib and Al-Soof, 2008; Ghomian et al., 2008; Imbus et al., 2006; Xie and Economides, 2009; Zeidouni et al., 2009). CO_2-EOR is a proven technology that certainly can offset, if not exceed, the costs of CO_2 capture, transportation, and injection of CO_2 storage projects.

However, the actual costs of CO_2 capture, compression, and transportation (if pipelines are not readily available) are too high, making CO_2-EOR economically unattractive. Adding to this the lack of a proper regulatory framework among other soft issues (e.g., public perception), we do not foresee an important increase in the number of projects implementing CO_2-EOR from anthropogenic sources in the near future.

Finally, acid gas coinjection of an H_2S and CO_2 mixture has been reported as an alternative to EOR applications in carbonate formations. Zama field (Canada), Tengiz field (Kazakhstan), and Harweel field (Oman) are a few examples of carbonate reservoirs with ongoing or planned sour or acid gas injection as an EOR strategy (Abou-Sayed et al., 2005; Longworth et al., 1996; Moritis, 2008; O'Dell et al., 2006).

8.3 OFFSHORE VERSUS ONSHORE EOR

Enhanced oil recovery in offshore fields is constrained not only by reservoir lithology, as was described earlier in this chapter, but also by surface facilities, environmental regulations, and the like (Bondor et al., 2005; Manrique, 2009). Therefore, its offshore applicability is limited compared to that for onshore fields. The main drainage strategy of offshore fields has been pressure maintenance by gas and water injection.

Figures 8.6 and 8.7 show the distribution of oil recovery projects in the North Sea (Awan et al., 2008; Jayasekera and Goodyear, 2002; Jethwa et al., 2000; Talukda and Instefjord, 2008) and EOR opportunities in offshore Malaysia (Hamdan et al., 2005; Nadeson et al., 2004; Samsudin et al., 2005; Selamal et al., 2008; Sudirman et al., 2007).

On the other hand, Figure 8.8 shows oil production in the Gulf of Mexico (GOM) in shallow and deep waters from 1990, including a forecast until the year 2013. GOM oil production is mainly supported by water and/or gas injection (Harun et al., 2008; Liu et al., 2008; MMS, 2005).

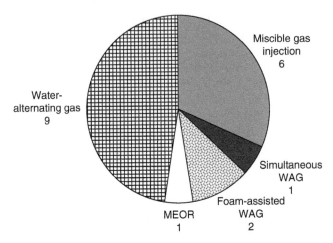

FIGURE 8.6 Recovery processes implemented in the North Sea. *Source: From Awan et al., 2008.*

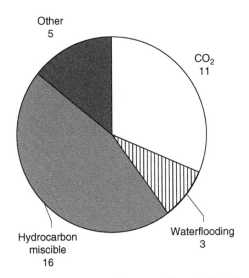

FIGURE 8.7 EOR opportunities in Malaysian offshore fields. *Source: From Samsudin et al., 2005.*

Despite environmental conditions in the area (e.g., hurricane seasons, another nontechnical issue), oil production from deep waters in GOM is expected to continue to increase in the future.

An increase in oil production will be associated with new projects, the announcement of numerous deepwater discoveries, improved reservoir characterization, recent drilling successes, and new federal incentives for the development of deep gas resources rather than EOR projects

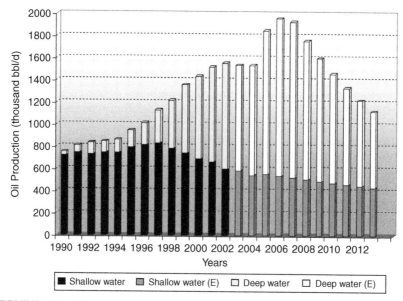

FIGURE 8.8 U.S. historic and forecasted oil production rates for GOM shallow and deepwater projects (E=estimated). *Source: From MMS, 2005.*

(Close et al., 2008; Iledare, 2008; Poll et al., 2009; Watson and Johnson, 2006). Mexico is another example of offshore reservoir production that is supported by gas injection, bottom water drive, and/or water injection. Regarding gas injection, it is important to note that Cantarell/Akal represents the largest N_2 injection project in the world (Cruz et al., 2009; Daltaban et al., 2009; Sánchez et al., 2005).

Similar to previous examples, most offshore environments are under continuous optimization strategies of both gas and waterflooding to extend field production life and to maximize oil recoveries (Awan et al., 2008; Beltrão et al., 2009; Daltaban et al., 2008; Huseby et al., 2008; Lindeloff et al., 2008; Monahan, 2009; Tealdi et al., 2008). Multiple initiatives, or EOR methods, have been tested—or have been proposed—for offshore fields, but only a few cases are listed here:

- CO_2-EOR storage has been proposed during the last decade (Awan et al., 2008; Gaspar et al., 2005; Hustad, 2009; Imbus et al., 2006; Xiang et al., 2008). CO_2-EOR from produced gas has been tested in Dulang field in Malaysia (Nadeson et al., 2004).
- High-pressure air injection has been proposed (e.g., Ekofisk, North Sea) but has not yet been tested for technical and economic reasons (Adetunji et al., 2005; Stokka et al., 2005). Mexico has considered the potential of air injection in offshore fields (Rodríguez and Christopher, 2004).

- Field applications of chemical EOR methods in offshore fields have not been widely documented in the literature.
- Foam-assisted WAG to improve gas mobility control based on a chemical method was successfully tested in Snorre field in the Norwegian North Sea (Awan et al., 2008; Blaker et al., 2002).
- Single-well alkali-surfactant-polymer combined with single-well partitioning tracers before and after ASP injection was successfully evaluated in Lagomar field at Maracaibo Lake in Venezuela (Hernandez, Alvarez, et al., 2002; Manrique et al., 2000).
- Single-well alkali-surfactant combined with single-well partitioning tracers before and after AS injection was successfully evaluated in Angsi field, offshore Terengganu in Malaysia (Othman et al., 2007).
- Polymer flooding has gained recent interest for offshore EOR applications, including the injection of colloidal dispersion gels (de Melo et al., 2005; Moritis, 2008; Spildo et al., 2009)

Although there are several initiatives to evaluate EOR potential in offshore fields, most of them are at the early stages of evaluation or may not be economically attractive with the current technology. Therefore, it is expected that commercial applications of EOR methods will likely not take place for at least a decade or two. Surface facility constraints and environmental regulations (e.g., chemical additives for EOR) also represent major hurdles for large EOR applications in offshore fields.

Offshore EOR projects are capital intensive. If we add the volatility of energy markets, the risk associated with this type of project is high, reducing the probability of its implementation or at least delaying it. Therefore, waterflooding and gas injection and their combined processes (e.g., WAG) will continue to support offshore production in the near term.

9

Closing Remarks

This book presented a comprehensive look at how business decisions lead to investments based on decision-making workflows. The field cases described demonstrate that decisions can and have been made using screening strategies without the need for more advanced techniques and time-consuming studies. This is not to say that advanced and more complex methods are unnecessary but that simpler engineering approaches often may be the right solution to decision making for enhanced oil recovery (EOR) projects before embarking on more costly and time-consuming exercises. Proper engineering judgment and physically sound analysis are key steps in this type of evaluation.

Such simplified approaches are followed as a matter of necessity and are not intended to substitute for conventional and rigorous numerical simulation and detailed reservoir studies. The less complicated workflows that are driven by limitations in data sources or/and short time frames are especially useful for small operators and independent companies that lack sufficient reservoir engineering resources.

Bias in screening can be mitigated by the use of several screening procedures, including those that rely on data mining. These procedures incorporate experience-based guidance as much as possible. Decision-making questions asked with the use of unproven technologies are good examples in which soft issues can play a significant role. The workflow presented here has been shown to be flexible enough for use in a number of different business settings, including EOR technologies that are not completely commercial.

Decision analysis can be enhanced by an open mind that accounts for most of the available possibilities. This can be greatly helped by a large knowledge base, which can be recorded in part through a well-organized database. Decision making can be phased according to time and/or data constraints with possible detrimental consequences.

We examined the important aspects of decision making, most of all in the earlier stages of EOR evaluations. Figure 9.1 shows what we hope was accomplished here. Our approach will certainly take you to Phase

Enhanced Oil Recovery
DOI: 10.1016/B978-1-85617-855-6.00015-2

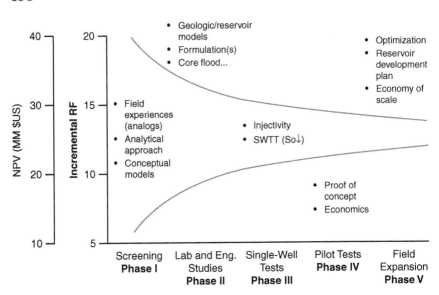

FIGURE 9.1   Uncertainty reduction in the phased project approach.

II and will help you establish an adequate mind-set for the later phases. In moving from Phase II to Phase IV, other references might be useful. The design of pilot tests requires other sources of information to answer the questions that need to be answered before committing resources for fieldwide deployment of EOR technologies. Enhanced and improved oil recovery technologies will continue to make strides. As other authors have said, we are not running out of oil; we are just running out of *easy* oil.

References

Aalund, L.R., 1988. Annual Production Report. Oil & Gas Journal, April 18.

Abdulraheem, A., Ahmed, M., Vantala, A., Parvez, T., 2009. Prediction of Rock Mechanical Parameters for Hydrocarbon Reservoirs Using Different Artificial Intelligence Techniques (SPE-126094). In: SPE Saudi Arabia Section Technical Symposium, Al-Khobar, May 9–11.

Abou-Sayed, A.S., Summers, C., Zaki, K., 2005. An Assessment of Engineering, Economical and Environmental Drivers of Sour-Gas Management by Injection (SPE-97628). In: SPE International Improved Oil Recovery Conference in Asia Pacific, Kuala Lumpur, December 5–6.

Adetunji, L.A., Teigland, R., Kleppe, J., 2005. Light-Oil Air-Injection Performance: Sensitivity to Critical Parameters (SPE-96844). In: SPE Annual Technical Conference and Exhibition, Dallas, October 9–12.

Adibhatia, B., Mohanty, K.K., 2007. Simulation of Surfactant-Aided Gravity Drainage in Fractured Carbonates (SPE-106601). In: SPE Reservoir Simulation Symposium, Houston, February 26–28.

Ahmed, T., 2006. Reservoir Engineering Handbook, third ed. Gulf Publishing.

Akram, F., 2008a. Effects of Well Placement using Multi-Segmented Wells in a Full Field Thermal Model for SAGD: Athabasca Oil Sands (Papers 2008-464). In: World Heavy Oil Congress. Edmonton, Alberta, March 10–12.

Akram, F., 2008b. Effects of Well Placement and Intelligent Completions on SAGD in a Full-Field Thermal-Numerical Model for Athabasca Oil Sands (SPE-117704). In: International Thermal Operations and Heavy Oil Symposium, Calgary, October 20–23, 2008.

Albahlani, A.M., Babadagli, T.A., 2008. Critical Review of the Status of SAGD: Where Are We and What Is Next? (SPE-113283). In: SPE Western Regional and Pacific Section AAPG Joint Meeting, Bakersfield, March 29–April 2.

Al-Bahar, M.A., Merril, R., Peake, W., Reza Oskui, M.J., 2004. Evaluation of IOR Potential within Kuwait (SPE-88716). In: 11th Abu Dhabi International Petroleum Exhibition and Conference, Abu Dhabi, October 10–13.

Alberta Chamber of Resources (ACR), 2004. Oil Sand Technology Roadmap: Unlocking the Potential. January 30. http://www.acr-alberta.com/.

Alberta Research Council (ARC), 2006. Prize^{TM}, Version 3.1 Manual. February.

Alberta Research Council (ARC), 2006. Prize^{TM}, Version 3.1. December.

Al-Dhafeeri, A.M., Nasr-El-Din, H.A., Al-Mubarak, H.K., Al-Ghamdi, J., 2008. Gas Shut-off Treatment in Oil Carbonate Reservoirs in Saudi Arabia (SPE-114323). In: SPE Annual Technical Conference and Exhibition, Denver, September 21–24.

Aldrich, H.S., Ashcraft Jr., T.L., Puerto, M.C., Reed, R.L., 1984. Oil Recovery Method Using Sulfonate Surfactants Derived from Extracted Aromatic Feedstocks. US Patent 4452708, June 5.

Alegre, L., Morokooka, C.K., Rocha, A.F., 1993. Intelligence Diagnosis of rod pumping problems (SPE-26516). In: SPE Annual Technical Conference, Houston, October 3–6.

Algharaib, M., Al-Soof, N.A., 2008. Economical Modeling of CO_2 Capturing and Storage Projects (SPE-120815). In: SPE Saudi Arabia Section Technical Symposium, Al-Khobar, May 10–12.

Algharaib, M., 2009. Potential Applications of CO_2-EOR in the Middle East (SPE-120231). In: SPE Middle East Oil and Gas Show and Conference, Bahrain, March 15–18.

Al-Hashim, H.S., Obiora, V., Al-Yousef, H.Y., Fernandez, F., Nofal, W., 1996. Alkaline Surfactant Polymer Formulation for Saudi Arabian Carbonate Reservoirs (SPE-35353). In: SPE/DOE Improved Oil Recovery Symposium, Tulsa, April 21–24.

Ali, J.K., 1994. Neural Networks: A New Tool for the Petroleum Industry? (SPE-27561). In: European Petroleum Computer Conference, Aberdeen, March 15–17.

Allain, O., Houze, O.P., 1992. A Practical artificial intelligence application in well testing interpretation (SPE-24287). In: SPE European Petroleum Conference, Stavanger, Norway, May 25–27.

Allan, J., Sun, S.Q., 2003. Controls on Recovery Factor in Fractured Reservoirs: Lessons Learned from 100 Fractured Fields (SPE-84590). In: SPE Annual Technical Conference and Exhibition, Denver, October 5–8.

Al-Taq, A.A., Nasr-El-Din, H.A., Beresky, J.K., Naimi, K.M., Sierra, L., Eoff, L., 2008. Simultaneous Acid Diversion and Water Control in Carbonate Reservoirs: A Case History From Saudi Arabia (SPE-106951). SPE Reservoir Evaluation & Engineering 11 (5), 882–891.

Alvarado, V., 2002. Analytical Simulation for Evaluation of EOR Opportunities in Venezuela: IFL Projects and Opportunity Maps. Master's Thesis in Exploration and Production, IFP, Paris.

Alvarado, V., Manrique, E., Vasquez, Y., Noreide, M., 2003. An approach for full-field EOR simulations based on fast evaluation tools. In: 24th Annual Workshop and Symposium for the IEA Collaborative Project on Enhanced Oil Recovery, September 7–10.

Alvarado, V., Ranson, A., Hernandez, K., Manrique, E., Matheus, J., Prosperi, N., Liscano, T., 2002. Selection of EOR/IOR Opportunities Based on Machine Learning (SPE-78332). In: 13th SPE European Petroleum Conference, Aberdeen, October 29–31.

Alvarado, V., Reich, E.M., Yi, Y., Postsch, K., 2006. Integration of a Risk-Management Tool and an Analytical Simulator for Assisted Decision-Making in IOR (SPE-100217). In: SPE EUROPEC/EAGE Annual Conference and Exhibition, Vienna, June 12–15.

Alvarado, V., Stirpe, M., La Roque, C., Ponce, R., Farías, M., 2002. Streamline Simulation for Enhanced Oil Recovery: Review and Laboratory Tests. INGEPET, EXPL-4-VA-84, Lima.

Alvarado, V., Thyne, G., Murrell, G.R., 2008. Screening Strategy for Chemical Enhanced Oil Recovery in Wyoming Basin (SPE-115940). In: SPE Annual Technical Conference and Exhibition, Denver, September 21–24.

Alvarez, C., Manrique, E., Alvarado, V., Saman, A., Surguchev, L., Eilertsen, T., 2001. WAG pilot at VLE field and IOR opportunities for mature fields at Maracaibo Lake (SPE-72099). In: SPE Asia Pacific Improved Oil Recovery Conference, Kuala Lumpur, October 6–9.

Alvarez, J.M., Sawatzky, R.P., Forster, L.M., Coates, R.M., 2008. Alberta's Bitumen Carbonate Reservoirs—Moving Forward with Advanced R&D (Paper 2008-467). World Heavy Oil Congress, Edmonton, March 10–12.

Antoniadi, D.G., Arzhanov, F.G., Garushev, A.R., Ishkhanov, V.G., 1993. Thermal Recovery Methods at the Former SU Oil Fields. Neft. Khoz 10, 24–25 28–29, October (in Russian).

Aoudia, M., Al-Kasimi, L.H.A., Al-Shibli, M.N.S., 2007. Evaluation of Alkyl Ether Sulfonates in Very High Salines Conditions: Application in EOR. In: 28th IEA Annual Workshop & Symposium Collaborative Project on EOR, Task B, Paper B-4, Vedbaek, Denmark, September 4–6.

Arihara, N., Yoneyama, T., Akita, Y., XiangGuo, L., 1999. Oil Recovery Mechanisms of Alkali-Surfactant-Polymer Flooding (SPE-54330). In: SPE Asia Pacific Oil and Gas Conference and Exhibition, Jakarta, April 20–22.

Armstrong, L., Edmunds, J.L., Clare, J.B., 2008. Environmental Considerations for Transferring Offshore Facilities to National Oil Company Operators (SPE-111704). In: SPE International Conference on Health, Safety, and Environment in Oil and Gas Exploration and Production, Nice, April 15–17.

Artola, F.A.V., Alvarado, V., 2006. Sensitivity Analysis of Gassmann's Fluid Substitution Equations: Some Implications in Feasibility Studies of Time-Lapse Seismic Reservoir Monitoring. Journal of Applied Geophysics 59 (1), 47–62.

Asgarpour, S., 2009. Hydrocarbons and Nuclear Energy: Optimization of Oil Sands Development from the Perspectives of Energy Supply, Environment and Economic Prosperity. Petroleum Technology Alliance Canada (PTAC), May.

Asghari, K., Dong, M., Shire, J., Coleridge, T., Nagrampa, J., Grassick, J., 2007. Development of a Correlation Between Performance of CO_2 Flooding and the Past Performance of Waterflooding in Weyburn Oil Field (SPE-99789). SPE Production & Operations 22 (2), 260–264.

Awan, A.R., Teigland, R., Kleppe, J.A., 2008. Survey of North Sea Enhanced-Oil-Recovery Projects Initiated During the Years 1975 to 2005 (SPE-99546). SPE Reservoir Evaluation & Engineering 11 (3), 497–512.

Babadagli, T., Sahin, S., Kalfa, U., Celebioglu, D., Karabakal, U., Topguder, N.N., 2008. Development of Heavy Oil Fractured Carbonate Bati Raman Field: Evaluation of Steam Injection Potential and Improving Ongoing CO_2 Injection (SPE-115400). In: SPE Annual Technical Conference and Exhibition, Denver, September 21–24.

Bachu, S., Brulotte, M., Grobe, M., Stewart, S., 2000. Suitability of the Alberta Subsurface for Carbon-Dioxide Sequestration in Geological Media. Earth Sciences Report 2000-11, Alberta Energy and Utilities Board, March.

Bagci, A.S., Gumrah, F., 2004. Effects of CO_2 and CH_4 Addition to Steam on Recovery of West Kozluca Heavy Oil (SPE-86953). In: SPE International Thermal Operations and Heavy Oil Symposium and Western Regional Meeting, Bakersfield, March 16–18.

Balch, R.S., Hart, D.M., Weiss, W.W., Broadhead, R.F., 2000. Regional Data Analysis to Better Predict Drilling Success: Brushy Canyon Formation, Delaware Basin, New Mexico (SPE-75145). In: SPE Improved Oil Recovery Symposium, Tulsa, April 13–17.

Barge, D., Al-Yami, F., Uphold, D., Zahedi, A., Deemer, A., Carreras, P.E., 2009. Steamflood Piloting the Wafra Field Eocene Reservoir in the Partitioned Neutral Zone, Between Saudi Arabia and Kuwait (SPE-120205). In: SPE Middle East Oil and Gas Show and Conference, Bahrain, March 15–18.

Bauer, M., 2006. Near-Term Opportunities and Emerging Economies: A view from Mexico. In: G8/IEA/CSFL Workshop on Near Term Opportunities for Carbon Capture and Storage, San Francisco, August 22–23.

Beaudette-Hodsman, C., Macleod, B., Venkatadri, R., 2008. Production of High Quality Water for Oil Sands Application (SPE-117840). In: International Thermal Operations and Heavy Oil Symposium, Calgary, October 20–23.

Beecroft, W.J., Mani, V., Wood, A.R.O., Rusinek, I., 1999. Evaluation of Depressurisation, Miller Field, North Sea (SPE-56692). In: SPE Annual Technical Conference and Exhibition, Houston, October 3–6.

Begg, S.H., Bratvold, R.B., Campbell, J.M., 2003. Shrinks Or Quants: Who Will Improve Decision-Making (SPE-84238). In: SPE Annual Technical Conference and Exhibition, Denver, October 5–8.

Belaifa, E., Tiab, D., Dehane, A., Jokhio, S., 2003. Effect of Gas Recycling on the Enhancement of Condensate Recovery in Toual Field Algeria, A Case Study (SPE-80899). In: SPE Production and Operations Symposium, Oklahoma City, March 22–25.

Bellarby, J., 2009. Well Completion Design. Elsevier.

Beltrão, R.L.C., Sombra, C.L., Lage, A.C.V.M., Fagundes Netto, J.R., Henriques, C.C.D., 2009. Challenges and New Technologies for the Development of the Pre-Salt Cluster, Santos Basin, Brazil (OTC-19880). In: Offshore Technology Conference, Houston, May 4–7.

Bersak, A.F., Kadak, A.C., 2007. Integration of Nuclear Energy with Oil Sands Projects for Reduced Greenhouse Gas Emissions and Natural Gas Consumption. Nuclear Energy and Sustainability Program, Massachusetts Institute of Technology, Report MIT-NES-TR-009, August. http://web.mit.edu/canes/publications/abstracts/nes/mit-nes-009.html.

Bibars, O.A., Hanafy, H.H., 2004. Waterflood Strategy—Challenges and Innovations (SPE-88774). In: 11th SPE Abu Dhabi International Petroleum Conference (ADIPEC), Abu Dhabi, October 10–13.

Bickel, J.E., Bratvold, R.B., 2007. Decision Making in the Oil and Gas Industry: From Blissful Ignorance to Uncertainty-Induced Confusion (SPE-109610). In: SPE Annual Technical Conference and Exhibition, Anaheim, November 11–14.

Biglarbigi, K., Paul, G.W., Ray, R.M. 1994. A Methodology for Prediction of Oil recovery by Infill Drilling (SPE/DOE 27761). In: SPE/DOE Ninth Symposium on Improved Oil Recovery, Tulsa, April 17–20.

Blaker, T., Aarra, M.G., Skauge, A., Rasmussen, L., Celius, H.K., Martinsen, H.A., Vassenden, F., 2002. Foam for Gas Mobility Control in the Snorre Field: The FAWAG Project (SPE-78824). SPE Reservoir Evaluation & Engineering 5 (4), 317–323.

Blanksby, J., Hicking, S., Milne, W., 2005. Deployment of High Horsepower ESPs to Extend Brent Field Life (SPE-96797). In: Offshore Europe Oil and Gas Exhibition, Aberdeen, September 6–9.

Bleakley, W.B., 1974. Production Report—Enhanced Oil Recovery. Oil & Gas Journal March 25.

Boge, R., Lien, S.K., Gjesdal, A., Hansen, A.G., 2005. Turning a North Sea Oil Giant into a Gas Field—Depressurization of the Statfjord Field (SPE-96403). In: Offshore Europe Oil and Gas Exhibition and Conference, Aberdeen, September 6–9.

Bondor, P.L., Hite, J.R., Avasthi, S.M., 2005. Planning EOR Projects in Offshore Oil Fields (SPE-94637). In: SPE Latin American and Caribbean Petroleum Engineering Conference, Rio de Janeiro, June 20–23.

Bortolotti, V., Macini, P., Srisuriyachai, F., 2009. Laboratory Evaluation of Alkali and Alkali-Surfactant-Polymer Flooding Combined with Intermittent Flow in Carbonatic Rocks (SPE-122499). In: Asia Pacific Oil and Gas Conference & Exhibition, Jakarta, August 4–6.

Bos, C.F.M., 2005. A Framework for Uncertainty Quantification and Technical-to-Business Integration for Improved Investment Decision-Making (SPE-94109). In: SPE EUROPEC/EAGE Annual Conference, pp. 13–16, Madrid, June.

Bradford, R.A., Compton, J.D., Hollis, P.R., 1980. Operational Problems in North Burbank Unit Surfactant/Polymer Project (SPE-7799). Journal of Petroleum Technology 32 (1), 11–17.

Brashear, J.P., Becker, A., Biglarbigi, K., Ray, R.M., 1989. Incentives, Technology, and EOR at Lower Oil Prices (SPE-17454). Journal of Petroleum Technology 41 (2), 164–170.

Brashear, J.P., Biglarbigi, K., Ray, M.R., 1991. Impact of Recent Federal Tax and R and D Initiatives on Enhanced Oil Recovery (SPE-22622). In: SPE Annual Technical Conference and Exhibition, Dallas, October 6–9.

Brashear, J.P., 1994. Improved Oil Recovery in the United States: The Race Between Technology Advancement and Resource Abandonment (SPE-27778). In: SPE/DOE Improved Oil Recovery Symposium, Tulsa, April 17–20.

Briones, M., Zambrano, J.A., Zerpa, C., 2002. Study of Gas-Condensate Well Productivity in Santa Barbara Field, Venezuela, by Well Test Analysis (SPE-77538). In: SPE Annual Technical Conference and Exhibition, San Antonio, September 29–October 2.

Bou-Mikael, S., Faisal Asmadi, F., Marwoto, D., Cease, C., 2000. Minas Surfactant Field Trial Tests Two Newly Designed Surfactants with High EOR Potential (SPE-64288). In: SPE Asia Pacific Oil and Gas Conference and Exhibition, Brisbane, October 16–18.

Bu, T., Soreide, I., Kydland, T., 1993. IOR Screening: What Went Wrong?. In: Seventh European IOR Symposium, Moscow, October 27–29.

Buckner, L.A., 1994. Tax Incentive for Enhanced Oil Recovery Projects (SPE-27781). In: SPE/DOE Improved Oil Recovery Symposium, Tulsa, April 17–20.

Butler, R.M., 1991. Thermal Recovery of Oil and Bitumen. Prentice Hall, 1991.

Button, P., Peterson, C., 2004. Enhanced Gravity Drainage Through Immiscible CO_2 Injection in the Yates Field (TX). In: 10th Annual CO_2 Flooding Conference, Midland, December.

Buza, J.W., 2008. An Overview of Heavy and Extra Heavy Oil Carbonate Reservoirs in the Middle East (IPTC-12426). In: International Petroleum Technology Conference, Kuala Lumpur, December 3–5.

Byfield, M., 2009. Husky and PTAC Propose A Game-Changing CO_2 EOR Project. Oil & Gas Inquirer. *http://www.oilandgasinquirer.com/columns/column.asp?article=magazine %2Fcolumns%2F090105%2FMAG_COL2009_J50000.html.*

Caers, K., Srinivasan, S., Journel, A.G., 2000. Geostatistical Quantification of Geological Information for a Fluvial-Type North Sea Reservoir (SPE-66310). SPE Reservoir Evaluation & Engineering 3 (5), 457–467.

Campanella, J.D., Wadleigh, E.E., Gilman, J.R., 2000. Flow Characterization – Critical for Efficiency of Field Operations and IOR (SPE-58996). In: SPE International Petroleum Conference and Exhibition, Villahermosa, February 1–3.

Carcoana, A., 1992. Applied Enhanced Oil Recovery. Prentice Hall.

Chakabaev, S.E., Ivanov, V.A., Shakhovoi, A.I., Tokarev, V.P., 1978. Injection of Water Thickened by Polyacrylamide—Industrial Experiment in the Karazhanbas Field. Geol. Nefti Gaza 1, 7–12 (in Russian).

Chang, H.L., Sui, X., Xiao, L., Guo, Z., Yao, Y., Xiao, Y., Chen, G., Song, K., Mack, J.C., 2006. Successful Field Pilot of In-Depth Colloidal Dispersion Gel (CDG) Technology in Daqing Oil Field (SPE-89460). SPE Reservoir Evaluation & Engineering 9 (6), 664–673.

Chang, H.L., Zhang, Z.Q., Wang, Q.M., Xu, Z.S., Guo, Z.D., Sun, H.Q., Cao, X.L., Qiao, Q., 2006. Advances in Polymer Flooding and Alkaline/Surfactant/Polymer Processes as Developed and Applied in the People's Republic of China (SPE-89175). Journal of Petroleum Technology 58 (2), 84–89.

Chaparral Energy Inc., 2009. Mid Continent, North Burbank Unit—Osage County, Oklahoma. *http://www.chaparralenergy.com/index.php?page=mid_continent* May 9.

Chattopadhyay, S.K., Ram, B., Bhattacharya, R.N., Das, T.K., 2004. Enhanced Oil Recovery by In-Situ Combustion Process in Santhal Field of Cambay Basin, Mehsana, Gujarat, India—A Case Study (SPE-89451). In: SPE/DOE Symposium on Improved Oil Recovery, Tulsa, April 17–21.

Chekani, M., Mackay, E.J., 2006. Impact on Scale Management of the Engineered Depressurization of Waterflooded Reservoirs: Risk Assessment Principles and Case Study (SPE-86472). SPE Production & Operations 21 (2), 174–181.

Chen, H.L., Lucas, L.R., Nogaret, L.A.D., Yang, H.D., Kenyon, D.E., 2000. Laboratory Monitoring of Surfactant Imbibition Using Computerized Tomography (SPE-59006). In: SPE International Petroleum Conference and Exhibition, Villahermosa, February 1–3.

Chen, H.L., Lucas, L.R., Nogaret, L.A.D., Yang, H.D., Kenyon, D.E., 2001. Laboratory Monitoring of Surfactant Imbibition Using Computerized Tomography (SPE-69197). SPE Reservoir Evaluation & Engineering 4 (1), 16–25.

Christensen, J.R., Stenby, E.H., Skauge, A., 2001. Review of WAG Field Experience (SPE-71203). SPE Reservoir Evaluation & Engineering 4 (2), 97–106.

Christensen, N.P. (Ed.), 2005. Report on the Current State and the Need for Further Research on CO_2 Capture and Storage. European Carbon Dioxide Network (CO_2 net), Rev. 6, August. *http://www.co2net.com/infocentre/reports/CCSRTDStrategyRev6.pdf.*

Chung, T.H.C., Carroll, H.B., Lindsey, R., 1995. Application of Fuzzy Expert Systems for EOR Project Risk Analysis (SPE-30741). In: SPE Annual Technical Conference and Exhibition, Dallas, October 22–25.

Civan, F., 2007. Reservoir formation damage: fundamentals, modeling, assessment, and mitigation, second ed. Elsevier.

Clampitt, R.L., Reid, T.B., 1975. An Economic Polymer Flood in the North Burbank Unit, Osage County, Oklahoma (SPE-5552). In: Fall Meeting of the Society of Petroleum Engineers of AIME, Dallas, September 28–October 1.

Clancy, J.P., Philcox, J.E., Watt, J., Gilchrist, R.E., 1985. Cases and economics for improved oil and gas recovery using nitrogen. In: 36th Petroleum Soc. & Canadian Soc. Petroleum Geologist Technical Mtg. Preprints, 85-36-2, Edmonton, June 2–5.

Clark Jr., R.A., Ludolph, B., 2003. Voidage Replacement Ratio Calculations in Retrograde Condensate to Volatile Oil Reservoirs Undergoing EOR Processes (SPE-84359). In: SPE Annual Technical Conference and Exhibition, pp. 5–8, Denver, October.

Close, F., McCavitt, B., Smith, B., 2008. Deepwater Gulf of Mexico Development Challenges Overview (SPE-113011). In: SPE North Africa Technical Conference & Exhibition, Marrakech, March 12–14.

Cokinos, J.S., Huff, B.G., Frailey, S.M., Seyler, B., Grube, J.P., 2004. Reservoir Management Using the Illinois State Geological Survey's Waterflood Database (SPE-91440). In: SPE Eastern Regional Meeting, Charleston, West Virginia, September 15–17.

Collins, P.M., Dusseault, M.B., Dorscher, D.E., Kueber, E., 2008. Implementing CHOPS in the Karazhanbas Heavy Oil Field, Kazakhstan (Paper 2008-500). World Heavy Oil Congress, Edmonton, March 10–12.

Cosentino, L., 2001. Integrated Reservoir Management. IFP Publications, Editions Technip, Paris, p. 310.

Costa, A.P.A., Schiozer, D.J., Moczydlower, P., Bedrikovetsky, P., 2008. Use of Representative Models to Improve the Decision Making Process of Chemical Flooding in a Mature Field (SPE-115442). In: SPE Russian Oil and Gas Technical Conference and Exhibition, Moscow, October 28–30.

Craig Jr., F.F., 1993. The Reservoir Engineering Aspects of Waterflooding. In: SPE of AIME, Dallas.

Crooks, N., 2008. Neuquén governor: Country to implement Oil Plus program—Argentina. New Sage Energy, Press Releases, September 24. http://www.newsage.ca/PressReleases/PressReleaseDetails/2008/NeuquengovernorCountrytoimplementOilPlusprogramArgentina/default.aspx.

Crude oil price, 2009. Refiner average domestic crude oil acquisition cost. Energy Information Administration. http://tonto.eia.doe.gov/dnav/pet/hist_xls/R1200____3m.xls, April.

Cruz, L., Sheridan, J., Aguirre, E., Celis, E., García-Hernandez, J., 2009. Relative Contribution To Fluid Flow From Natural Fractures in the Cantarell Field (SPE-122182). In: Latin American and Caribbean Petroleum Engineering Conference, Cartagena de Indias, May 31–June 3.

Cunha, J.C., 2007. Importance of Economic and Risk Analysis on Today's Petroleum Engineering Education (SPE-109638). In: SPE Annual Technical Conference and Exhibition, Anaheim, November 11–14.

Da Cruz, P.S., Horne, R.N., Deutsch, C.V., 2004. The Quality Map: A Tool for Reservoir Uncertainty Quantification and Decision Making (SPE-87642). In: SPEREE, February.

Dakota Gasification Company, 2009. CO_2 Capture and Storage, August. http://www.dakotagas.com/CO2_Capture_and_Storage/index.html.

Daltaban, T.S., Lozada, A.M., Villavicencio, P.A., Torres, F.M., 2008. Managing Water and Gas Production Problems in Cantarell: A Giant Carbonate Reservoir in Gulf of Mexico (SPE-117233). In: Abu Dhabi International Petroleum Exhibition and Conference, Abu Dhabi, November 3–6.

Das, S., 2008. Electro-Magnetic Heating in Viscous Oil Reservoir (SPE-117693). In: International Thermal Operations and Heavy Oil Symposium, Calgary, October 20–23.

Da Silva, I.P.G., de Melo, M.A., Luvizotto, J.M., Lucas, E.F., 2007. Polymer Flooding: A Sustainable Enhanced Oil Recovery in the Current Scenario (SPE-107727). In: Latin American & Caribbean Petroleum Engineering Conference, Buenos Aires, April 15–18.

De Haan, H.J., Van Lookeren, J., 1969. Early Results of the First Large-Scale Steam Soak Project in the Tia Juana Field, Western Venezuela (SPE-1913). Journal of Petroleum Technology 21 (1), 101–110.

de Melo, M.A., da Silva, I.P.G., de Godoy, G.M.R., Sanmartim, A.N., 2002. Polymer Injection Projects in Brazil: Dimensioning, Field Application and Evaluation (SPE-75194). In: SPE/DOE Improved Oil Recovery Symposium, Tulsa, April 13–17.

de Melo, M.A., de Holleben, C.R.C., da Silva, I.P.G., de Barros, C.A., da Silva, G.A., Rosa, A.J., Lins, A.G., de Lima, J.C., 2005. Evaluation of Polymer Injection Projects in Brazil (SPE-94898). In: SPE Latin American and Caribbean Petroleum Engineering Conference, Rio de Janeiro, June 20–23.

de Souza, J.C., da S. Cursino, D. F., de O. Pádua, K. G., 2005. Twenty Years of Steam Injection in Heavy-Oil Fields (SPE-94808). In: SPE Latin American and Caribbean Petroleum Engineering Conference, Rio de Janeiro, June 20–23.

Dehghani, K., Ehrlich, R., 1998. Evaluation of steam injection process in light oil reservoirs (SPE-49016). In: SPE Annual Tech. Conference, New Orleans, September 27–30.

Delamaide, E., Corlay, P., Wang, D., 1994. Daqing Oil Field: The Success of Two Pilots Initiates First Extension of Polymer Injection in a Giant Oil Field (SPE-27819). In: SPE/DOE Improved Oil Recovery Symposium, Tulsa, April 17–20.

Dell, J.J., Meakin, S., Cramwinckel, J., 2008. Sustainable Water Management in the Oil and Gas Industry: Use of the WBCSD Global Water Tool to Map Risks (SPE-111960). In: SPE International Conference on Health, Safety, and Environment in Oil and Gas Exploration and Production, Nice, April 15–17.

Delshad, M., Najafabadi, N.F., Sepehrnoori, K., 2009. Scale Up Methodology for Wettability Modification in Fractured Carbonates (SPE-118915). In: SPE Reservoir Simulation Symposium, The Woodlands, Texas, February 2–4.

Demin, W., Jiecheng, C., Junzheng, W., Zhenyu, Y., Yuming, Y., Hongfu, L., 1999. Summary of ASP Pilots in Daqing Oil Field (SPE-57288). In: SPE Asia Pacific IOR Conference, Kuala Lumpur, October 25–26.

Department of the Treasury, Internal Revenue Service (IRS), 2007. Tier II Issue—Enhanced Oil Recovery Credit Directive #1. Washington, D.C., May 2. *http://www.irs.gov/businesses/article/0id=169940,00.html*.

Dezen, F., Morooka, C., 2001. Field Development Decision Making Under Uncertainty: A Real Option Valuation Approach (SPE-69595). In: SPE Latin American and Caribbean Petroleum Engineering Conference, Buenos Aires, March 25–28.

Diaz, D., Somaruga, C., Norman, C., Romero, J., 2008. Colloidal Dispersion Gels Improve Oil Recovery in a Heterogeneous Argentina Waterflood (SPE-113320). In: SPE/DOE Symposium on Improved Oil Recovery, Tulsa, April 20–23.

Ding, E., Harrison, P., Dozzo, J., Lin, C.Y., 2009. Prudhoe Bay: Rebuilding a Giant Oil and Gas Full Field Model (SPE-120967). In: SPE Western Regional Meeting, San Jose, March 24–26.

Dinnie, N.C., Fletcher, A.J.P., Finch, J.H., 2002. Strategic Decision Making in the Upstream Oil and Gas Industry: Exploring Intuition and Analysis (SPE-77910). In: SPE Asia Pacific Oil and Gas Conference and Exhibition, Melbourne, October 8–10.

Dino, R., Rocha, P.S., Sanches, C., Le Thiez, P., 2007. EOR and Storage Activities driven by CO_2 in Brazil—Experience from the Buracica and Miranga oil fields performance: Planned Operations. In: 2nd International Symposium: Capture and Geological Storage of CO_2, Paris, October 3–5.

Doleschall, S., Szittar, A., Udvardi, G., 1992. Review of the 30 Years' Experience of the CO_2 Imported Oil Recovery Projects in Hungary (SPE-22362). In: International Meeting on Petroleum Engineering, Beijing, March 24–27.

Domitrović, D., Sunjerga, S., Jelić-Balta, J., 2004. Numerical Simulation of Tertiary CO_2 Injection at Ivanić Oil Field, Croatia (SPE-89361). In: SPE/DOE Symposium on Improved Oil Recovery, Tulsa, April 17–21.

Donaldson, A.B., 1997. Reflections on a Downhole Steam Generator Program (SPE-38276). In: SPE Western Regional Meeting, Long Beach, June 25–27.

Doraiah, A., Ray, S., Gupta, P., 2007. In-Situ Combustion Technique to Enhance Heavy-Oil Recovery at Mehsana, ONGC—A Success Story (SPE-105248). In: SPE Middle East Oil and Gas Show and Conference, Kingdom of Bahrain, March 11–14.

Drummond, A., Fishlock, T., Naylor, P., Rothkopf, B., 2001. An Evaluation of Post-Water-flood Depressurisation of the South Brae Field, North Sea (SPE-71487). In: SPE Annual Technical Conference and Exhibition, New Orleans, September 30–October 3.

Dudfield, C.F.O., 1988. Evaluating Options in Field Development Planning. Offshore 48 (11), 41–43.

Duiveman, M.W., Herwin, H., Grivot, P., 2005. Integrated Management of Water, Lean Gas and Air Injection: The Successful Ingredients to IOR Projects on the Mature Handil Field (SPE-93858). In: SPE Asia Pacific Oil and Gas Conference and Exhibition, Jakarta, April 5–7.

Edwards, K.A., Anderson, B., Reavie, B., 2002. Horizontal Injectors Rejuvenate Mature Miscible Flood—South Swan Hills Field (SPE-77302). SPE Reservoir Evaluation & Engineering 5 (2), 174–182.

Elemo, R.O., Elmtalab, J.A., 1993. Practical Artificial Intelligence Application in EOR Projects (SPE-26248). SPE Computer Applications 4 (2), 17–21.

Elliot, K.T., Kovscek, A.R., 2001. A Numerical Analysis of the Single-Well Steam Assisted Gravity Drainage (SW-SAGD) Process (SUPRI TR-124). DOE Topical Report, June.

El Mahdi, A., Ayoub, M., Daif-Allah, S., Negahban, S., Ribeiro, M., Bahamaish, J., 2007. An Integrated Approach for a Full-Field, Optimized Development of a Carbonate Reservoir on a Shallow Marine Environment, Abu Dhabi (SPE-105416). In: SPE Middle East Oil and Gas Show and Conference, Kingdom of Bahrain, March 11–14.

Ernandez, J., 2009. EOR Projects in Venezuela: Past and Future. In: ACI *Optimising EOR Strategy 2009*, London, March 11–12.

Eson, R.L., 1982. Downhole Steam Generator—Field Tests (SPE-10745). In: SPE California Regional Meeting, San Francisco, March 24–26.

Fabel, G., Neunhoeffer, T., Rudschinski, D., Sasse, J., Scheer, T., 1999. Reservoir Management of Mature Oil Fields by Integrated Field Development Planning (SPE-54114). In: SPE Thermal Operation & Heavy Oil International Symposium, Bakersfield, March 17–19.

Fattahi, S., Youn, S., Zhang, W., Wightman, D.M., Shang, R., 2004. Reservoir Heterogeneity and its Effect on SAGD (Steam-Assisted Gravity Drainage) Oil Production from the McMurray Formation, Christina Lake Area, Alberta. In: CSPG-Canadian Heavy Oil Association–CWLS Joint Conference, Calgary, May 31–June 4.

Ferrel, H.H., Conley, D., Casad, B.M., Stokke, O.M., 1980. Polymer flood filtration improvement. US Patent 4212748, July 15.

Figueiredo Jr., F.P., Branco, C.C.M., Prais, F., Salomão, M.C., Mezzomo, C.C., 2007. The Challenges of Developing a Deep Offshore Heavy-Oil Field in Campos Basin (SPE-107387). In: SPE Latin American & Caribbean Petroleum Engineering Conference, Buenos Aires, April.

Flanery, S.O., McCarty, S.C., 2008. Recent Legal Developments in Carbon Sequestration (SPE-116231). In: SPE Eastern Regional/AAPG Eastern Section Joint Meeting, Pittsburgh, October 11–15.

Fletcher, A.J.P., Morrison, G.R., Developing a Chemical, E.O.R., 2008. Pilot Strategy for a Complex, Low Permeability Water Flood (SPE-112793). In: SPE/DOE Symposium on Improved Oil Recovery, Tulsa, April 20–23.

Frampton, H., Morgan, J.C., Cheung, S.K., Munson, L., Chang, K.T., Williams, D., 2004. Development of a Novel Waterflood Conformance Control System (SPE-89391). In: SPE/DOE Symposium on Improved Oil Recovery, Tulsa, April 17–21.

Freedenthal, C., Taylor, M.A., 1989. Impact of Natural Gas Market Factors on Developing Resources (SPE-19662). In: SPE Annual Technical Conference and Exhibition, San Antonio, October 8–11.

French, T., 1999. Evaluation of the Sho-Vel-Tum Alkali-Surfactant-Polymer (ASP) Oil Recovery Project—Stephens County. OK: Final Report (November 19, 1998–May 18, 1999), U.S. DOE Fossil Energy Report No. DOE/SW/45030-1. July.

Fuller, S.M., Sarem, A.M., Gould, T.L., 1992. Screening Waterfloods for Infill Drilling Opportunities (SPE-22333). In: Society of Petroleum Engineers.

Gachuz-Muro, H., Sellami, H., 2009. Analogous Reservoirs to Chicontepec, Alternatives of Exploitation for this Mexican Oil Field (SPE-120265). In: EUROPEC/EAGE Conference and Exhibition, Amsterdam, June 8–11.

Gael, B.T., Gross, S.J., McNaboe, G.J., 1995. Development Planning and Reservoir Management in the Duri Steam Flood (SPE-29668). In: 65th SPE Western Regional Meeting Bakersfield, March 8–10.

Gallagher, J.J., Kemshell, D.M., Taylor, S.R., Mitro, R.J., 1999. Brent Field Depressurization Management (SPE-56973). In: Offshore Europe Oil and Gas Exhibition and Conference, Aberdeen, September 7–10.

Garbutt, C.F., 1997. Innovative Treating Processes Allow Steamflooding with Poor Quality Oilfield Water (SPE-38799). In: SPE Annual Technical Conference and Exhibition, San Antonio, October 5–8.

Garcia, M.C., Chiaravallo, N., Sulbaran, A., El Chiriti, K., Chirinos, A., 2001. Production Restarting on Asphaltene-Plugged Oil Wells in a Lake Maracaibo Reservoir (SPE-69513). In: SPE Latin American and Caribbean Petroleum Engineering Conference, Buenos Aires, March 25–28.

Garland, E., 2005. Environmental Regulatory Framework in Europe: An Update (SPE-93796). In: SPE/EPA/DOE Exploration and Production Environmental Conference, Galveston, March 7–9.

Garland, E., 2007. From "Forget It, It's Illegal," to "Yes, but …": The Saga of Carbon Underground Storage Regulatory Framework (SPE-109254). In: Offshore Europe Oil and Gas Exhibition and Conference, Aberdeen, September 4–7.

Garnier, O.F., Cupcic, F., 2002. Alternate Fuels for Thermal Projects (SPE-78958). In: SPE International Thermal Operations and Heavy Oil Symposium and International Horizontal Well Technology Conference, Calgary, November 4–7.

Gaspar, A.T.F.S., Suslick, S.B., Ferreira, D.F., Lima, G.A.C., 2005. Enhanced Oil Recovery With CO_2 Sequestration: A Feasibility Study of a Brazilian Mature Oil Field (SPE-94939). In: SPE/EPA/DOE Exploration and Production Environmental Conference, Galveston, March 7–9.

Gaspar, A.T.F.S., Suslick, S.B., Ferreira, D.F., Lima, G.A.C., 2005. Economic Evaluation of Oil Production Project with EOR: CO_2 Sequestration in Depleted Oil Field (SPE-94922). In: SPE Latin American and Caribbean Petroleum Engineering Conference, Rio de Janeiro, June 20–23.

Geddes, C.J., Curlett, H.B., 2005. Leveraging a New Energy Source to Enhance Heavy-Oil and Oil-Sands Production (SPE-97781). In: SPE/PS-CIM/CHOA International Thermal Operations and Heavy Oil Symposium, Calgary, November 1–3.

Gharbi, R.B.C., 2000. An Expert System for Selecting and Designing EOR Processes. Journal of Petroleum Science and Engineering 27 (1), 33–47.

Ghomian, Y., Urun, M., Pope, G.A., Sepehrnoori, K., 2008. Investigation of Economic Incentives for CO_2 Sequestration (SPE-116717). In: SPE Annual Technical Conference and Exhibition, Denver, September 21–24.

Gillham, T.H., Cerveny, B.W., Fornea, M.A., Bassiouni, Z., 1998. Low Cost IOR: An Update on the W. Hackberry Air Injection Project (SPE-39642). In: SPE/DOE Improved Oil Recovery Symposium, Tulsa, April 19–22.

Gillham, T.H., Cerveny, B.W., Turek, E.A., Yannimaras, D.V., 1997. Keys to Increasing Production via Air Injection in Gulf Coast Light Oil Reservoirs (SPE-38848). In: SPE Annual Technical Conference and Exhibition, San Antonio, October 5–8.

Godec, M.L., Kosowski, B., Haverkamp, D.S., Hochheiser, B., 1993. The Potential Role of Future Environmental Regulations on the Domestic Petroleum Industry (SPE-25833). In: SPE Hydrocarbon Economics and Evaluation Symposium, Dallas, March 29–30.

Godec, M.L., 2009. Environmental Performance of the Exploration and Production Industry: Past, Present, and Future (SPE-120918). In: SPE Americas E&P Environmental and Safety Conference, San Antonio, March 23–25.

Gogarty, W.B., 1976. Status of Surfactant or Micellar Methods (SPE-5559). Journal of Petroleum Technology 28 (1), 93–102.

Gogarty, W.B., 1978. Micellar/Polymer Flooding An Overview (SPE-7041). Journal of Petroleum Technology 30 (8), 1089–1101.

Gomes, J.S., Ribeiro, M.T., Fouchard, P., Twombley, B.N., Negahban, S., Al-Baker, S., 2002. Geological Modeling of a Tight Carbonate Reservoir for Improved Reservoir Management of a Miscible WAG Injection Project (SPE-78529). In: Abu Dhabi International Petroleum Exhibition and Conference, Abu Dhabi, October 13–16.

Good, W.K., Scott, J.D., Luhning, R.W., 1994. Review and Assessment of Steam Assisted Gravity Drainage (SAGD) Applications in Canada. In: 14th World Petroleum Congress, Stavanger, Norway, May 29–June 1.

Goodyear, S.G., Gregory, A.T., 1994. Risk Assessment and Management in IOR Projects (SPE-28844). In: European Petroleum Conference, London, October 25–27.

Govind, P.A., Das, S., Srinivasan, S., Wheeler, T.J., 2008. Expanding Solvent SAGD in Heavy Oil Reservoirs (SPE-117571). In: International Thermal Operations and Heavy Oil Symposium, Calgary, October 20–23.

Greaves, M., Xia, T.X., Ayasse, C., 2005. Underground Upgrading of Heavy Oil Using THAI— "Toe-to-Heel Air Injection" (SPE-97728). In: SPE/PS-CIM/CHOA International Thermal Operations and Heavy Oil Symposium, Calgary, November 1–3.

Green, D.W., Willhite, G.P., 1998. Enhanced Oil Recovery. Society of Petroleum Engineers, Richardson. Texas, p. 545.

Griffith, J.O.H., Cox, T.F., 1986. The Economics of Late-Life Field Production in the North Sea (SPE-15863). In: European Petroleum Conference, London, October 20–22.

Grills, T.L., Vandal, B., Hallum, F., Trost, P., 2002. Case History: Horizontal Well SAGD Technology Is Successfully Applied to Produce Oil at LAK Ranch in Newcastle, Wyoming (SPE-78964). In: SPE International Thermal Operations and Heavy Oil Symposium and International Horizontal Well Technology Conference, Calgary, November 4–7.

Gu, H., Yang, R., Guo, S., Guan, W., Yue, X., Pan, Q., 1998. Study on Reservoir Engineering: ASP (Alkali, Surfactant, Polymer) Flooding Pilot Test in Karamay Oilfield (SPE-50918). In: Sixth SPE Oil & Gas International Conference, Beijing, November 2–6.

Guan, L., McVay, D.A., Jensen, J.L., Voneiff, G.W., 2002. Evaluation of a Statistical Infill Candidate Selection Technique (SPE-75718-MS). In: SPE Gas Technology Symposium, Calgary, April 30–May 2.

Guedes, S., 2008. 70% Recovery Factor: Petrobras Perspective. In: Presentation at Rio Oil & Gas Expo and Conference, Rio de Janeiro, September 15–18.

Guerillot, D.R., 1988. EOR Screening with an Expert System (SPE-17791). In: Petroleum Computer Conference, San Jose, June 27–29, 1988.

Gutiérrez, D., Miller, R.J., Taylor, A.R., Thies, B.P., Kumar, V.K., 2008b. Buffalo Field High-Pressure Air Injection Projects: Technical Performance and Operational Challenges (SPE-113254). In: SPE/DOE Symposium on Improved Oil Recovery, Tulsa, April 20–23.

Gutiérrez, D., Taylor, A.R., Kumar, V.K., Ursenbach, M.G., Moore, R.G., Mehta, S.A., 2008a. Recovery Factors in High-Pressure Air Injection Projects Revisited (SPE-108429). SPE Reservoir Evaluation & Engineering 11 (6), 1097–1106.

Hamada, G.M., Elshafei, M.A., 2009. Neural Network Prediction of Porosity and Permeability of Heterogeneous Gas Sand Reservoir (SPE-126042). In: SPE Saudi Arabia Section Technical Symposium, Al-Khobar, Saudi Arabia, May 9–11.

Hamdan, M.K., Darman, N., Hussain, D., Ibrahim, Z., 2005. Enhanced Oil Recovery in Malaysia: Making It a Reality (SPE-93329). In: SPE Asia Pacific Oil and Gas Conference and Exhibition, Jakarta, April 5–7.

Hamilton, G., 2009. Strategies for CO_2 Reductions in the U.S. Enhanced Oil Recovery Production Segment: Utilizing Energy Efficiency as the First Step (SPE-121003). In: SPE Americas E&P Environmental and Safety Conference, San Antonio, March 23–25.

Han, M., Xiang, W., Zhang, J., Jiang, W., Sun, F., 2006. Application of EOR Technology by Means of Polymer Flooding in Bohai Oil Fields (SPE-104432). In: SPE International Oil & Gas Conference and Exhibition, Beijing, December 5–7.

Han, D.K., Yang, C.Z., Zhang, Z.Q., Lou, Z.H., Chang, Y.I., 1999. Recent development of enhanced oil recovery in China. Journal of Petroleum Science and Engineering 22, 181–188.

Hanzlik, E.J., 2003. Forty Years of Steam Injection in California—The Evolution of Heat Management (SPE-84848). In: SPE International Improved Oil Recovery Conference in Asia Pacific, Kuala Lumpur, October 20–21.

Harun, A.F., Blanchard, T.J., Erdogmus, M., 2008. Managing Hydrate Risks for a Black Oil Long Subsea Tie-Back When Water Cut Predictions Exceed Original Design (SPE-111138). In: SPE North Africa Technical Conference & Exhibition, Marrakech, March 12–14.

Hascakir, B., Acar, C., Demiral, B., Akin, S., 2008. Microwave Assisted Gravity Drainage of Heavy Oils (IPTC-12536). In: International Petroleum Technology Conference, Kuala Lumpur, December 3–5.

Hascakir, B., Babadagli, T., Akin, S., 2008. Experimental and Numerical Modeling of Heavy-Oil Recovery by Electrical Heating (SPE-117669). In: International Thermal Operations and Heavy Oil Symposium, Calgary, October 20–23.

Haugen, Å., Fernø, M.A., Graue, A., 2008. Numerical Simulation and Sensitivity Analysis of In-Situ Fluid Flow in MRI Laboratory Waterfloods of Fractured Carbonate Rocks at Different Wettabilities (SPE-116145). In: SPE Annual Technical Conference and Exhibition, Denver, September 21–24.

Heffern, E.W., Suffridge, F.E., Valcho, J.J., Fenoglio, D.J., 1982. Development of a Crude Oil Sulfonate for Micellar/Polymer Flooding of the Salt Creek Field in Natrona County, Wyoming (SPE-9813). Journal of Petroleum Technology 34 (10), 2283–2290.

Henry, J.D., 1977. The Effect of Oil Price Policy on Tertiary Oil Recovery (SPE-6354). In: SPE Economics and Evaluation Symposium, Dallas, February 21–26.

Henson, R.M., 2001. Geologically Based Screening Criteria for Improved Oil Recovery Projects. Ph.D. Dissertation. Heriot Watt University, Department of Petroleum Engineering, June.

Henson, R., Todd, A., Corbett, P., 2002. Geologically Based Screening Criteria for Improved Oil Recovery Projects (SPE-75148). In: SPE/DOE IOR Symposium, Tulsa, April 13–17.

Hernández, C., Alvarez, C., Saman, A., De Jongh, A., de Audemard, N., 2002. Monitoring WAG Pilot at VLE Field, Maracaibo Lake, by Perfluorocarbon and Fluorined Benzoic Acids Tracers (SPE-75259). In: SPE/DOE Improved Oil Recovery Symposium, Tulsa, April 13–17.

Hernández, C., Chacón, L.J., Anselmi, L., Baldonedo, A., Qi, J., Dowling, P.C., Pitts, M.J., 2001. ASP System Design for an Offshore Application in the La Salina Field, Lake Maracaibo (SPE-69544). In: SPE Latin American and Caribbean Petroleum Engineering Conference, Buenos Aires, March 25–28.

Hernández, C., Chacon, L.J., Anselmi, L., Baldonedo, A., Qi, J., Dowling, P.C., Pitts, M.J., 2003. ASP System Design for an Offshore Application in La Salina Field, Lake Maracaibo (SPE-84775). SPE Reservoir Evaluation & Engineering 6 (3), 147–156.

Hernández, C., Chacon, L., Anselmi, L., Angulo, R., Manrique, E., Romero, E., de Audemard, N., Carlisle, C., 2002. Single Well Chemical Tracer Test to Determine ASP (Alkaline, Surfactant, Polymer) Injection Efficiency at Lagomar VLA–6/9/21 Area, C4 Member, Lake Maracaibo, Venezuela (SPE-75122). In: 13th SPE/DOE Improved Oil Recovery Symposium, Tulsa, April 13–17.

Hernández, C., Chacón, L., Moreno, R., Anselmi, L., Manrique, E., 2002b. The State of Improved Oil Recovery by ASP Injection in PDVSA. In: International Energy Agency 23rd Annual Workshop & Symposium: Collaborative Project on Enhanced Oil Recovery, September 8–11.

Hernández, K.Y., Liscano, T., Ranson, A., Manrique, E., Matheus, J., Alvarado, V., 2002. Use of Machine Learning for EOR Method Selection in Venezuelan Reservoirs. In: IEA Collaborative Project on Enhanced Oil Recovery, 23rd International Workshop and Symposium, Caracas, September 8–11.

Herrera, L., Mendoza, H., 2001. Controllability and Observability Issues for SAGD Intelligent Control in Venezuela (SPE-69699). In: SPE International Thermal Operations and Heavy Oil Symposium, Porlamar, Margarita Island, Venezuela, March 12–14.

Hirasaki, G.J., Miller, C.A., Pope, G.A., 2006. Surfactant Based Enhanced Oil Recovery and Foam Mobility Control—3rd Annual & Final Technical Report. Rice University and U.S. Department of Energy, Report No. DE-FC26-03NT15406, July.

Hodgin, J.E., Harrell, D.R., 2006. The Selection, Application, and Misapplication of Reservoir Analogs for the Estimation of Petroleum Reserves (SPE-102505). In: SPE Annual Technical Conference and Exhibition, San Antonio, September 24–27.

Holliday, G.H., 2009. Comments Regarding the USEPA Proposed Class VI Well Rule (SPE-120774). In: SPE Americas E&P Environmental and Safety Conference, San Antonio, March 23–25.

Holtz, M.H., 2008. Summary of Sandstone Gulf Coast CO_2 EOR Flooding Application and Response (SPE-113368). In: SPE/DOE Symposium on Improved Oil Recovery, Tulsa, April 20–23.

Hongfu, L., Guangzhi, L., Peihui, H., Zhenyu, Y., Xiaolin, W., Chen Guangyu, C., Dianping, X., Peiqiang, J., 2003. Alkaline/Surfactant/Polymer (ASP) Commercial Flooding Test in Central Xing2 Area of Daqing Oilfield (SPE-84896). In: SPE Asia Pacific International Improved Oil Recovery Conference, (IIORC03), Kuala Lumpur, October 20–21.

Hongmin, Y., Baoquan, Y., Guorui, X., Jiexiang, W., Shao Ran Ren, L., Weimin, L., Liang, X., Haitao, G., 2008. Air Foam Injection for IOR: From Laboratory to Field Implementation in ZhongYuan Oilfield China (SPE-113913). In: SPE/DOE Symposium on Improved Oil Recovery, Tulsa, April 20–23.

Horne, E., Sun, S., Nichols, L., Agatonovic, V., 2002. An Alternative Water Supply for Heavy Oil Steam Generation Plants: Potential Sources of Brackish Water From the Cretaceous Mannville Group in the Cold Lake Region (SPE-79003). In: SPE International Thermal Operations and Heavy Oil Symposium and International Horizontal Well Technology Conference, Calgary, November 4–7.

Hudson, J.W., Jochen, J.E., Jochen, V.A., 2000. Practical Technique to Identify Infill Potential in Low-Permeability Gas Reservoirs Applied to the Milk River Formation in Canada (SPE-59779). In: SPE/CERI Gas Technology Symposium, Calgary, Canada, April 3–5.

Hudson, J.W., Jochen, J.E., Spivey, J.P., 2001. Potential Methods to High-Grade Infill Opportunities Applied to the Mesaverde, Morrow and Cotton Valley Formations (SPE-68598). In: SPE Hydrocarbon Economics and Evaluation Symposium, Dallas, April 2–3.

Hughes, B.L., Sarma, H.K., 2006. Burning Reserves For Greater Recovery? Air Injection Potential In Australian Light Oil Reservoirs (SPE-101099). In: SPE Asia Pacific Oil & Gas Conference and Exhibition, Adelaide, September 11–13.

Huseby, O., Andersen, M., Svorstol, I., Dugstad, O., 2008. Improved Understanding of Reservoir Fluid Dynamics in the North Sea Snorre Field by Combining Tracers, 4D Seismic, and Production Data (SPE-105288). SPE Reservoir Evaluation & Engineering 11 (4), 768–777.

Hustad, C.W., 2009. Capturing, Managing and Gathering CO_2 for EOR Onshore and Offshore: Challenges and Opportunities. In: ACI Optimising EOR Strategy Conference, London, March 11–12.

Hustad, C.W., Austell, J.M., 2004. Mechanisms and Incentives to Promote the Use and Storage of CO_2 in the North Sea. In: Roggenkamp, M.M., Hammer, U. (Eds.), European Energy Law Report I, pp. 355–380, Intersentia.

Hutchin, L.A., Burton, R.K., Macintosh, D.J., 1996. An Expert System for Analyzing Well Performance (SPE-35705). In: SPE Western Regional Meeting, Anchorage, May 22–24.

Ibatullin, R.R., Ibragimov, N.G., Khisamov, R.S., Podymov, E.D., Shutov, A.A., 2002. Application and method based on artificial intelligence for selection of structures and Screening of technologies for enhanced oil recovery (SPE-75175). In: SPE/DOE IOR Symposium, Tulsa, April 13–17.

Iledare, O.O., 2008. Profitability of Deepwater Petroleum Leases: Emipirical Evidence From the U.S. Gulf of Mexico Offshore Region (SPE-116602). In: SPE Annual Technical Conference and Exhibition, Denver, September 21–24.

Iledare, O.O., 2004. Analyzing the Impact of Petroleum Fiscal Arrangements and Contract Terms on Petroleum E&P Economics and the Host Government Take (SPE-88969). In: SPE Annual International Conference and Exhibition, Abuja, Nigeria, August 2–4.

Imbus, S., Orr, F.M., Kuuskraa, V.A., Kheshgi, H., Bennaceur, K., Gupta, N., et al., 2006. Critical Issues in CO_2 Capture and Storage: Findings of the SPE Advanced Technology Workshop (ATW) on Carbon Sequestration (SPE-102968). In: SPE Annual Technical Conference and Exhibition, San Antonio, September 24–27.

IRIS Research, 2007. SWORD Analytical Tool Manual (Version 2.1). October.

Islam, M.R., Wadadar, S.S., Bansal, A., 1991. Enhanced Oil Recovery of Ugnu Tar Sands of Alaska Using Electromagnetic Heating with Horizontal wells (SPE-22177). In: International Arctic Technology Conference, Anchorage, May 29–31.

Jayasekera, A.J., Goodyear, S.G., 2002. Improved Hydrocarbon Recovery in the United Kingdom Continental Shelf: Past, Present and Future (SPE-75171). In: SPE/DOE Improved Oil Recovery Symposium, Tulsa, April 13–17.

Jelgersma, F., 2007. Redevelopment of the Abandoned Dutch Onshore Schoonebeek Oilfield With Gravity Assisted Steam Flooding (IPTC-11700). In: International Petroleum Technology Conference, Dubai, December 4–6.

Jensen, J.L., Currie, I.D.A., 1990. New Method for Estimating the Dykstra-Parsons Coefficient To Characterize Reservoir Heterogeneity (SPE-17364). SPE Reservoir Engineering 5 (3), 369–374.

Jethwa, D.J., Rothkopf, B.W., Paulson, C.I., 2000. Successful Miscible Gas Injection in a Mature U.K. North Sea Field (SPE-62990). In: SPE Annual Technical Conference and Exhibition, Dallas, October 1–4.

Jian, H., Wenfen, H., 2006. Novel Approach To Predict Potentiality of Enhanced Oil Recovery (SPE-99261). In: Intelligent Energy Conference and Exhibition, Amsterdam, April 11–13.

Jimenez, J., 2008. The Field Performance of SAGD Projects in Canada (IPTC-12860). In: International Petroleum Technology Conference, Kuala Lumpur, December 3–5.

Joseph, J., Taber, F., David, M., Seright, R.S., 1996. EOR Screening Criteria Revisited (SPE-35385). In: SPE/DOE Improved Oil Recovery Symposium, Tulsa, April 21–24.

Kabir, C.S., Jamaluddin, A.K.M., 2002. Asphaltene Characterization and Mitigation in South Kuwait's Marrat Reservoir (SPE-80285). SPE Production & Facilities 17 (4), 251–258.

Karaoguz, O.K., Topguder, N.N., Lane, R.H., Kalfa, U., Celebioglu, D., 2007. Improved Sweep in Bati Raman Heavy-Oil CO_2 Flood: Bullhead Flowing Gel Treatments Plug Natural Fractures (SPE-89400). SPE Reservoir Evaluation & Engineering 10 (2), 164–175.

Kaura, J., Sierra, J., 2008. Successful Field Application in Continuous DTS Monitoring under Harsh Environment of SAGD Wells Using Improved Optical Fiber Technology (SPE-117206). In: SPE International Thermal Operations and Heavy Oil Symposium, Calgary, October 20–23.

Keplinger, C.H., 1965. Economic Considerations Affecting Steam Flood Prospects (SPE-1097). In: SPE Symposium on Petroleum Economics and Evaluation, Dallas, February 8–9.

Kleinschmidt, R.F., Lorenz, P.B., 1976. North Burbank Unit Tertiary Recovery Pilot Test. Phillips Petroleum Company, Annual report, May.

Knapp, R.M., Yang, X., 1999. Identifying Opportunities to Increase Oil Recovery from Fluvial-dominated Deltaic Reservoirs in Oklahoma (SPE-52226). In: SPE Mid-Continent Operations Symposium, Oklahoma City, March 28–31.

Koning, E.J.L., Mentzer, E., Heemskerk, J., 1988. Evaluation of a Pilot Polymer Flood in the Marmul Field, Oman (SPE-18092). In: SPE Annual Technical Conference and Exhibition, Houston, October 2–5.

Krawchuk, P., Beshry, M.A., Brown, G.A., Brough, B., 2006. Predicting the Flow Distribution on Total E&P Canada's Joslyn Project Horizontal SAGD Producing Wells Using Permanently Installed Fiber-Optic Monitoring (SPE-102159). In: SPE Annual Technical Conference and Exhibition, San Antonio, September 24–27.

Kulkarni, M.M., Chen, H.L., Brummert, A.C., 2008. CO$_2$ IOR Evaluation for the U.S. Rocky Mountain Assets (SPE-113297). In: SPE/DOE Symposium on Improved Oil Recovery, Tulsa, April 20–23.

Kumar, A., Al-Ajmi, M.F., Al-Anzi, E., Clark, R.A., Khater, M., Lantz, J., 2007. Water Shut off Techniques to Combat Premature Water Breakthrough in Mauddud Carbonate Reservoir—An Efficacy Analysis (IPTC-11713). In: International Petroleum Technology Conference, Dubai, December 4–6.

Kumar, V.K., Gutierrez, D., Moore, R.G., Mehta, S.A., 2008. Air Injection and Waterflood Performance Comparison of Two Adjacent Units in the Buffalo Field (SPE-104479). SPE Reservoir Evaluation & Engineering 11 (5), 848–857.

Lacerda, G.M., Patriota, J.H., Pereira, J.I., Torres, J.S., De Lima, L.A., 2008. Alto do Rodrigues GeDIg Pilot—Case Study for Continuous Steam Injection Recovery Combined with Real-Time Operation (SPE-112242). In: SPE Intelligent Energy Conference and Exhibition, Amsterdam, February 25–27.

Lake, L.W., 1989. Enhanced Oil Recovery. Prentice-Hall, p. 600.

Laughton, D., 2003. Effects of Input Price Uncertainty on Asset Valuation (SPE-84236). In: SPE Annual Technical Conference and Exhibition, Denver, October 5–8.

Lawrence, J.J., Maer, N.K., Stern, D., Corwin, D., Idol, L.W., Jay, W.K., 2002. Nitrogen Tertiary Recovery Study: Managing a Mature Field (SPE-78527). In: SPE International Petroleum Exhibition and Conference, Abu Dhabi, October 13–16.

Lawry, T.F.A., 1974. Review of Tertiary Recovery in Illinois (SPE-4869). In: SPE Midwest Oil and Gas Industry Symposium, Indianapolis, March 28–29.

Lazo Lazo, J.G., Pacheco, M.A.C., Vellasco, M.M.V.R., 2007. Hybridizing Genetic Algorithms and Real Options to Approach of the Optimal Decision Rule for Oilfield Development Under Uncertainties (SPE-107937). In: SPE Latin American & Caribbean Petroleum Engineering Conference, Buenos Aires, April 15–18.

Leaute, R.P., 2002. Liquid Addition to Steam for Enhancing Recovery (LASER) of Bitumen with CSS: Evolution of Technology from Research Concept to a Field Pilot at Cold Lake (SPE-79011). In: SPE International Thermal Operations and Heavy Oil Symposium and International Horizontal Well Technology Conference, Calgary, November 4–7.

Leonard, J., 1984. Annual Production Report—Enhanced Oil Recovery. Oil & Gas Journal April 2.

Leonard, J., 1982. Annual Production Report. Oil & Gas Journal, April 5.

Leonard, J., 1986. Production/Enhanced Oil Recovery Report. Oil & Gas Journal, April 24.

Lerat, O., Adjemian, F., Auvinet, A., Baroni, A., Bemer, E., Eschard, R., et al., 2009. Modelling of 4D Seismic Data for the Monitoring of the Steam Chamber Growth During SAGD Process (Paper CIPC 2009-095). In: Canadian International Petroleum Conference, Calgary, June 16–18.

Levine, S., Sigmon, R., Douglas, S., 2002. Yates Field—Super Giant of the Permian Basin. Houston Geological Society Bulletin 45 (3), 39–45, 47–49, 51.

Levitt, D.B., Jackson, A.C., Heinson, C., Britton, L.N., Malik, T., Dwarakanath, V., Pope, G.A., 2009. Identification and Evaluation of High-Performance EOR Surfactants (SPE-100089-PA). SPE Reservoir Evaluation & Engineering 12 (2), 243–253.

Li, J., Li, T., Yan, J., Zuo, X., Zheng, Y., Yang, F., 2009. Silicon Containing Scale Forming Characteristics and How Scaling Impacts Sucker Rod Pump in ASP Flooding (SPE-122966). In: SPE Asia Pacific Oil and Gas Conference & Exhibition, Jakarta, August 4–6.

Li, D., Shi, M., Wang, D., Li, Z., 2009. Chromatographic Separation of Chemicals in Alkaline Surfactant Polymer Flooding in Reservoir Rocks in the Daqing Oil Field (SPE-121598). SPE International Symposium on Oilfield Chemistry The Woodlands, Texas, April 20–22.

Lindeloff, N., Mogensen, K., van Lingen, P., Do, S.H., Frank, S., Noman, R., 2008. Fluid-Phase Behaviour for a Miscible-Gas-Injection EOR Project in a Giant Offshore Oil Field With Large Compositional Variations (SPE-115970). In: SPE Annual Technical Conference and Exhibition, Denver, September 21–24.

Lino, U., de, R.A., 2005. Case History of Breaking a Paradigm: Improvement of an Immiscible Gas-Injection Project in Buracica Field by Water Injection at the Gas/Oil Contact (SPE-94978). In: SPE Latin American and Caribbean Petroleum Engineering Conference, Rio de Janeiro, June 20–23.

Li-qiang, Y., Da-sheng, Z., Yu-huan, S., 2006. SAGD as Follow-Up to Cyclic Steam Stimulation in a Medium Deep and Extra Heavy-Oil Reservoir (SPE-104406). In: SPE International Oil & Gas Conference and Exhibition, Beijing, December 5–7.

Liu, B., Dessenberger, R., McMillen, K., Lach, J., Kelkar, M., 2008. Water-Flooding Incremental Oil Recovery Study in Middle Miocene to Paleocene Reservoirs, Deep-Water Gulf of Mexico (SPE-115669). In: SPE Asia Pacific Oil and Gas Conference and Exhibition, Perth, October 20–22.

Long, R.E., Nuar, M.F.A., 1982. Study of Getty Oil Co.'s Successful In-Situ Combustion Project in the Bellevue Field (SPE-10708). In: SPE Enhanced Oil Recovery Symposium, Tulsa, April 4–7.

Longworth, H.L., Dunn, G.C., Semchuck, M., 1996. Underground Disposal of Acid Gas in Alberta, Canada: Regulatory Concerns and Case Histories (SPE-35584). In: SPE Gas Technology Symposium, Calgary, April 28–May 1.

Lorenz, P.B., 1986. Postflood Evaluation of the North Burbank Surfactant-Polymer Pilot (NIPER-94). National Institute for Petroleum and Energy, Bartlesville, Oklahoma, June.

Louie, J., 2009. Operating Experience from CO_2 Miscible Floods Provides Design Guidelines for CO_2 Sequestration. Hatch Energy Report, February.

Lowry, P.H., Ferrell, H.H., Dauben, D.L., 1986. A Review and Statistical Analysis of Micellar-Polymer Field Test Data. National Petroleum Technology Office (NPTO). U.S. Department of Energy, Tulsa, Report No. DOE/BC/10830-4, November.

Macaulay, R.C., Krafft, J.M., Hartemink, M., Escovedo, B., 1995. Design of a Steam Pilot in a Fractured Carbonate Reservoir—Qarn Alam Field, Oman (SPE-30300). In: SPE International Heavy Oil Symposium, Calgary, June 19–21.

Machedon, V., Popescu, T., Paduraru, R., 1995. U.S. DOE Fossil Energy Report No. NIPER/BDM-0086 and Conf-940450, February, Romania; 30 Years of Experience in In Situ Combustion, pp. 83–96. U.S. DOE Field Application of In Situ Combustion—Past Performance/Future Applications Symposium, Tulsa, April 21–22.

Mack, J., 2005. Special Report: Upstream China—In-Depth Gel Shows Promise in Daqing IOR (Improved Oil Recovery) Pilot. Oil & Gas Journal 103 (26), 44–47.

Mackie, S.I., Welsh, M.B., 2006. An Oil and Gas Decision-Making Taxonomy. In: Paper SPE 100699 presented at the SPE Asia Pacific Oil & Gas Conference and Exhibition, Adelaide, September 11–13.

Mahinpey, N., Ambalae, A., Asghari, K., 2007. In Situ Combustion in Enhanced Oil Recovery (EOR). A Review. Chemical Engineering Communications 194 (8), 995–1021.

Maldal, T., Gilje, E., Kristensen, R., Karstad, T., Nordbotten, A., Schilling, B.E.R., et al., 1998. Planning and Development of Polymer Assisted Flooding for the Gullfaks Field Norway (SPE-35378). SPE Reservoir Evaluation & Engineering 1 (2), 161–168.

Malik, Q.M., Islam, M.R., 2000. CO_2 Injection in the Weyburn Field of Canada: Optimization of Enhanced Oil Recovery and Greenhouse Gas Storage with Horizontal Wells (SPE-59237). In: SPE/DOE Improved Oil Recovery Symposium, Tulsa, April 3–5.

Mamedov, Y.G., Bokserman, A.A., 1992. Application of Improved Oil Recovery in the Soviet Union (SPE-24162). In: Eighth SPE/DOE Enhanced Oil Recovery Symposium, Tulsa, April 22–24.

Manning, R.K., Pope, G.A., Lake, L.W., Paul, G.W., Wesson, T.C., 1983. A Technical Survey of Polymer Flooding Projects. Department of Energy, Report DOE/BC/10327-19, September.

Manrique, E., 2009. Enhanced Oil Recovery (EOR): Trends, Risks, and Rewards. In: ACI Optimising EOR Strategy, London, March 11–12.

Manrique, E., Araya, A., 2008. Capture and Geologic Sequestration—What It Is and Its Implications. Coal Preparation Society of America (CPSA) Journal 6 (4).

Manrique, E., Calderon, G., Mayo, L., Stirpe, M.T., 1998. Water-Alternating-Gas Flooding in Venezuela: Selection of Candidates Based on Screening Criteria of International Field Experiences (SPE-50645). In: SPE European Petroleum Conference, The Hague, October 20–22.

Manrique, E., de Carvajal, G., Anselmi, L., Romero, C., Chacon, L., 2000. Alkali/Surfactant/Polymer at VLA 6/9/21 Field in Maracaibo Lake: Experimental Results and Pilot Project Design (SPE-59363). In: SPE/DOE Improved Oil Recovery Symposium, Tulsa, April 2–5.

Manrique, E., Izadi, M., Kitchen, C., Alvarado, V., 2008. Effective EOR Decision Strategies with Limited Data: Field Cases Demonstration (SPE-113269). In: SPE/DOE Symposium on Improved Oil Recovery, Tulsa, April 20–23.

Manrique, E.J., Muci, V.E., Gurfinkel, M.E., 2007. EOR Field Experiences in Carbonate Reservoirs in the United States (SPE-100063). SPE Reservoir Evaluation & Engineering 10 (6), 667–686.

Manrique, E., Padron, R., Surguchev, L., De Mena, J., McKenna, K., 2000b. VLE WAG Injection Laboratory Field in Maracaibo Lake (SPE-65128). In: SPE European Petroleum Conference, Paris, October 24–25, 2000.

Manrique, E., Pereira, C., 2007. Identifying Viable EOR Thermal Processes in Canadian Tar Sands (Paper CIPC 2007-176). In: Eighth Canadian International Petroleum Conference (58th Annual Technical Meeting), Calgary, June 12–14.

Manrique, E., Ranson, A., Alvarado, V., 2003. Perspectives of CO_2 injection in Venezuela. In: 24th Annual Workshop and Symposium for the IEA Collaborative Project on Enhanced Oil Recovery in Canada, September 7–10.

Manrique, E., Wright, J.D., 2006. Screening Methods Help Operators Identify Viable EOR Opportunities. The American Oil & Gas Reporter, June.

Marinello, S.A., Lyon, F.L., Ballantine, W.T., 2001. Disposal of E & P Waste by Injection: An Assessment of Technology and Regulatory Impacts (SPE-66521). In: SPE/EPA/DOE Exploration and Production Environmental Conference, San Antonio, February 26–28.

Matheny, S.L., 1980. Production Report—Enhanced Oil Recovery. Oil & Gas Journal, March 31.

Mbaba, P.E., Caballero, E.P., 1983. Field Application of an Additive Containing Sodium Metasilicate During Steam Stimulation (SPE-12058). In: SPE Annual Technical Conference and Exhibition, San Francisco, October 5–8.

McCormack, M., 2001. Mapping of the McMurray Formation for SAGD. Journal of Canadian Petroleum Technology 40 (8), 21–28.

McMillen, S.J., 2004. International E&P Environmental Regulations: What Makes Sense for Our Industry? (SPE-86636). In: SPE International Conference on Health, Safety, and Environment in Oil and Gas Exploration and Production, Calgary, March 29–31.

Mendez, Z., Alvarez, J.M., Escobar, E., Colonomos, P., Campos, E., 1992. Cyclic Steam Injection with Additives: Laboratory and Field Test Results of Steam/Foam and Steam/Solvent Processes (SPE-24632). In: SPE Annual Technical Conference and Exhibition, Washington, October 4–7.

Mendoza, H.A., Finol, J.J., Butler, R.M., 1999. SAGD, Pilot Test in Venezuela (SPE-53687). In: SPE Latin American and Caribbean Petroleum Engineering Conference, Caracas, April 21–23.

Meyer, D.R., 1988. The Water Quality Control Station: A New Monitoring Tool for Injection Water (SPE-17524). In: SPE Rocky Mountain Regional Meeting, Casper, May 11–13.

Meyers, J.J., Pitts, M.J., Wyatt, K., 1992. Alkaline-Surfactant-Polymer Flood of the West Kiehl, Minnelusa Unit (SPE-24144). In: SPE/DOE Enhanced Oil Recovery Symposium, Tulsa, April 22–24.

Mezzomo, R.F., Luvizotto, J.M., Palagi, C.L., 2001. Improved Oil Recovery in Carmópolis Field: R&D and Field Implementations (SPE-69811). In: SPE Reservoir Evaluation & Engineering 4 (1), 4–10.

Mijnssen, F.C.J., Davies, A.H., Grondin, K., Keating, J., Hsu, C.F., Amthor, J., 2003. Bringing Al Huwaisah's Volume to Value (SPE-84285). In: SPE Annual Technical Conference and Exhibition, Denver, October 5–8.

Miller, J.A., 2008. How Processes Will Contribute to the Next Trillion Barrels, Journal of Petroleum Technology, 18–20, August.

Minami, K., Almeida, A.S., Diniz, M.A.C., Palagi, C.L., Assayag, M.I., 2003. Campos Basin: Challenges Still to Overcome (OTC-15225). In: Offshore Technology Conference, Houston, May 5–8.

Mineral Management Services (MMS), 2005. Reserves vs. Secondary Recovery in GOM: Summary data, May.

Mitchell, T., 1997. Machine Learning. McGraw-Hill.

Moffitt, P.D., Mitchell, J.F., 1983. North Burbank Unit Commercial Scale Polymerflood Project—Osage County, Oklahoma (SPE-11560). In: SPE Production Operations Symposium, Tulsa, February 27–March 1.

Mohaghegh, S., 2000. Virtual-Intelligence Applications in Petroleum Engineering: Part 1—Artificial Neural Networks (SPE-58046). Journal of Petroleum Technology 52 (9), 64–73.

Mohammed-Singh, L.J., Singhal, A.K., 2004. Lessons From Trinidad's CO_2 Immiscible Pilot Projects 1973–2003 (SPE-89364). In: SPE/DOE Symposium on Improved Oil Recovery, Tulsa, April 17–21.

Mohan, H., Carolus, M., Biglarbigi, K., 2008. The Potential for Additional Carbon Dioxide Flooding Projects in the United States (SPE-113975). In: SPE/DOE Symposium on Improved Oil Recovery, Tulsa, April 20–23.

Mohanty, K.K., 2006. Dilute Surfactant Methods for Carbonate Formations: Final Report. University of Houston and U.S. Department of Energy Report No. DE-FC26-02NT 15322, February.

Monahan, P., 2009. Flow Assurance Challenges for Offshore Deepwater in Republic of Congo (OTC-20279). In: Offshore Technology Conference, Houston, May 4–7.

Moon, T., 2008. Digital Technologies for the Next Trillion Barrels. Journal of Petroleum Technology, 58–63, September.

Moore, R.G., Mehta, S.A., Ursenbach, M.G., 2002. A Guide to High Pressure Air Injection (HPAI) Based Oil Recovery (SPE-75207). In: SPE/DOE Improved Oil Recovery Symposium, Tulsa, April 13–17.

Morel, D., Labastie, A., Jouenne, S., Nahas, E., 2007. Feasibility Study for EOR by Polymer Injection. In Deep Offshore Fields (IPTC-11800). In: International Petroleum Technology Conference, Dubai, December 4–6.

Moritis, G., 1990. Annual Production Report. Oil & Gas Journal, April 23.

Moritis, G., 1992. Annual Production Report. Oil & Gas Journal, April 20.

Moritis, G., 1994. EOR Production Report. Oil & Gas Journal, September 26.

Moritis, G., 1996. Oil Production Report. Oil & Gas Journal, April 14.

Moritis, G., 1998. EOR Oil Production Up Slightly. Report on Enhanced Oil Recovery. Oil & Gas Journal, April 18.

Moritis, G., 2000. Report on Enhanced Oil Recovery. Oil & Gas Journal, March 20.

Moritis, G., 2001. New Companies, Infrastructure, Projects Reshape Landscape for CO_2 EOR in the US. Oil & Gas Journal Special Report, May 14.

Moritis, G., 2002. Report on Enhanced Oil Recovery. Oil & Gas Journal, April 15.

Moritis, G., 2004. Report on Enhanced Oil Recovery. Oil & Gas Journal, April 12.

Moritis, G., 2006. Report on Enhanced Oil Recovery. Oil & Gas Journal, April 17.

Moritis, G., 2008. Special Report EOR/Heavy Oil Survey: More US EOR Projects Start but EOR Production Continues Decline. Oil & Gas Journal, April 21.

MSN Money, 2009. Energy: Light Sweet Crude Oil. U.S. Commodities. August 21. *http://moneycentral.msn.com/investor/market/commodities.aspx*.

Mungan, N., 2000. Enhanced Oil Recovery with High Pressure Nitrogen Injection (SPE-62547). In: SPE/AAPG Western Regional Meeting, Long Beach, June 19–22.

Muro, H.G., Campos, S.B., Alcazar, L.O., Rodríguez, J.A., Quebrache, A., 2007. Natural CO_2 Reservoir: A New Source for EOR Projects in Mexico (SPE-107445). In: SPE Latin American & Caribbean Petroleum Engineering Conference, Buenos Aires, April 15–18.

Muruaga, E., Flores, M., Norman, C., Romero, J., 2008. Combining Bulk Gels and Colloidal Dispersion Gels for Improved Volumetric Sweep Efficiency in a Mature Waterflood (SPE-113334). In: SPE/DOE Symposium on Improved Oil Recovery, Tulsa, April 20–23.

Mustafiz, S., Islam, M.R., 2008. State-of-the-Art Petroleum Reservoir Simulation. Pet. Sci. Technol. 26, 1303–1329.

Nadeson, G., Anua, N.A.B., Singhal, A., Ibrahim, R.B., 2004. Water-Alternating-Gas (WAG) Pilot Implementation, A First EOR Development Project in Dulang Field, Offshore Peninsular Malaysia (SPE-88499). In: SPE Asia Pacific Oil and Gas Conference and Exhibition, Perth, October 18–20.

Najafabadi, N.F., Delshad, M., Sepehrnoori, K., Nguyen, Q.P., Zhang, J., 2008. Chemical Flooding of Fractured Carbonates Using Wettability Modifiers (SPE-113369). In: SPE/DOE Symposium on Improved Oil Recovery, Tulsa, April 20–23.

Nakamura, S., Sarma, H.K., Umucu, T., Issever, K., Kanemitsu, M., 1995. A Critical Evaluation of a Steamflood Pilot in a Deep Heavy Carbonate Reservoir in Ikiztepe Field, Turkey (SPE-30727). In: SPE Annual Technical Conference and Exhibition, Dallas, October 22–25.

National Energy Technology Laboratory, 1983. Crude Oil Analysis Database. Retrieved November 28, 2007, from the National Energy Technology Laboratory Web site. *http://www.netl.doe.gov/technologies/oil-gas/Software/database.html*.

National Energy Technology Laboratory, 1984. NPC Public Database (NPCPUBDB.GEO). Retrieved November 28, 2007, from the National Energy Technology Laboratory Web site. *http://www.netl.doe.gov/technologies/oil-gas/Software/database.html*.

National Energy Technology Laboratory, 2005. Rocky Mountain Basins Produced Water Database. Retrieved November 28, 2007, from the National Energy Technology Laboratory Web site. *http://www.netl.doe.gov/technologies/oil-gas/Software/database.html*.

Needham, R.B., Doe, P.H., 1987. Polymer Flooding Review (SPE-17140). Journal of Petroleum Technology 39 (12), 1503–1507.

Neff, J.M., Hagemann, R., 2007. Environmental Challenges of Heavy Crude Oils: Management of Liquid Wastes (SPE-101973). In: SPE E & P Environmental and Safety Conference, Galveston, March 5–7.

Negrescu, M., 2008. Economic Modeling of an Oil and Gas Project Involving Carbon Capture and Storage: Snohvit LNG Field (Barents Sea, Norway) (SPE-107430). SPE Projects, Facilities & Construction 3 (2), 1–15.

NIPER, 1986. Enhanced Oil Recovery Information (revised edition), April.

Noran, D., 1976. Production Report. Oil & Gas Journal, April 5.

Noran, D., 1978. Annual Production Issue. Oil & Gas Journal, March 27.

Nouri, A., Vaziri, H., Belhaj, H., Islam, R., 2003. Effect of Volumetric Failure on Sand Production in Oil-Wellbores. (SPE-80448). In: SPE Asia Pacific Oil and Gas Conference and Exhibition, Jakarta, September 9–11.

Novosel, D., 2005. Initial Results of WAG CO_2 IOR Pilot Project Implementation in Croatia (SPE-97639). In: SPE International Improved Oil Recovery Conference in Asia Pacific, Kuala Lumpur, December 5–6.

Novosel, D., 2009. Thermodynamic Criteria and Final Results of WAG CO_2 Injection in a Pilot Project in Croatia (SPE-119747). In: SPE Middle East Oil and Gas Show and Conference, Bahrain, March 15–18.

Nummedal, D., Towler, B., Mason, C., Allen, M., 2003. Enhanced Oil Recovery in Wyoming—Prospects and Challenges. University of Wyoming.

O'Dell, M., Soek, H., van Rossem, S., 2006. Achieving the Vision in the Harweel Cluster, South Oman (SPE-102389). In: SPE Annual Technical Conference and Exhibition, San Antonio, September 24–27.

Ohms, D., McLeod, J., Graff, C.J., Frampton, H., Morgan, J.C., Cheung, S., Yancey, K., Chang, K.T., 2009. Incremental Oil Success From Waterflood Sweep Improvement in Alaska (SPE-121761). In: SPE International Symposium on Oilfield Chemistry, The Woodlands, Texas, April 20–22.

Onishi, T., Katoh, K., Takabayashi, K., Uematsu, H., Okatsu, K., Wada, Y., Ogata, Y., 2007. High Pressure Air Injection into Light Oil Reservoirs: Experimental Study on Artificial Ignition. In: IEA Collaborative Project on Enhanced Oil Recovery 28th Annual Workshop and Symposium, Denmark, September 4–7.

Othman, M., Chong, M.O., Sai, R.M., Zainal, S., Zakaria, M.S., Yaacob, A.A., 2007. Meeting the Challenges in Alkaline Surfactant Pilot Project Implementation at Angsi Field, Offshore Malaysia (SPE-109033). In: Offshore Europe Oil and Gas Exhibition and Conference, Aberdeen, September 4–7.

Ovalles, C., Vallejos, C., Vasquez, T., Martinis, J., Perez-Perez, A., Cotte, E., Castellanos, L., Rodriguez, H., 2001. Extra-Heavy Crude Oil Downhole Upgrading Process Using Hydrogen Donors under Steam Injection Conditions (SPE-69692). In: SPE International Thermal Operations and Heavy Oil Symposium, Porlamar, Venezuela, March 12–14.

Palmgren, C., Renard, G., 1995. Screening Criteria for the Application of Steam Injection and Horizontal Wells. In: Eighth European Symposium on Improved Oil Recovery, Vienna, May 15–17, 1995.

Panait-Paticaf, A., Ažerban, D., Ilie, N., 2006. Suplacu de Barcau Field—A Case History of a Successfull In-Situ Combustion Exploitation (SPE-100346). In: SPE EUROPEC/EAGE Annual Conference and Exhibition, Vienna, June 12–15.

Panda, M.N., Ambrose, G., Beuhler, G., McGuire, P.L., 2009. Optimized EOR, Design for the Eileen West End Area, Greater Prudhoe Bay (SPE-123030). SPE Reservoir Evaluation & Engineering 12 (1), 25–32.

Pandey, A., Beliveau, D., Suresh Kumar, M., Pitts, M.J., Qi, J., 2008. Evaluation of Chemical Flood Potential for Mangala Field, Rajasthan, India—Laboratory Experiment Design and Results (IPTC-12636). In: International Petroleum Technology Conference, Kuala Lumpur, December 3–5.

Pandey, A., Beliveau, D., Corbishley, D.W., Kumar, M.S., 2008. Design of an ASP Pilot for the Mangala Field: Laboratory Evaluations and Simulation Studies (SPE-113131). In: SPE Indian Oil and Gas Technical Conference and Exhibition, Mumbai, March 4–6.

Parmar, G., Zhao, L., Graham, J., 2009. Start-up of SAGD Wells: History Match, Wellbore Design and Operation. Journal of Canadian Petroleum Technology 48 (1).

Patil, S., Dandekar, A.Y., Patil, S.L., Khataniar, S., 2008. Low Salinity Brine Injection for EOR on Alaska North Slope (ANS) (IPTC-12004). In: International Petroleum Technology Conference, Kuala Lumpur, December 3–5.

Peden, J.M., Tovar, J.J., 1991. Sand prediction and Exclusion Decision Support Using an Expert System (SPE-23165). In: SPE Offshore European Conference, Aberdeen, September 3–6.

Pedenaud, P., Dang, F.A., 2008. New Water Treatment Scheme for Thermal Development: The SIBE Process (SPE-117561). In: SPE International Thermal Operations and Heavy Oil Symposium, Calgary, October 20–23.

Pedersen, F.B., Hanssen, T.H., Aasheim, T.I., 2006. How Far Can a State-of-the-Art NPV Model Take You in Decision Making? (SPE-99627). In: SPE EUROPEC/EAGE Annual Conference and Exhibition, Vienna, June 12–15.

Penney, R., Baqi Al Lawati, S., Hinai, R., Van Ravesteijn, O., Rawnsley, K., et al., 2007. First Full Field Steam Injection in a Fractured Carbonate at Qarn Alam, Oman (SPE-105406). In: SPE Middle East Oil and Gas Show and Conference, Bahrain, March 11–14.

Penney, R., Moosa, R., Shahin, G., Hadhrami, F., Kok, A., Engen, G., Van Ravesteijn, O., Rawnsley, K., Kharusi, B., 2005. Steam Injection in Fractured Carbonate Reservoirs: Starting a New Trend in EOR (IPTC-10727). In: International Petroleum Technology Conference, Doha, Qatar, November 21–23.

Perez-Perez, A., Gamboa, M., Ovalles, C., Manrique, E., 2001. Benchmarking of Steamflood Field Projects in Light/Medium Crude Oils (SPE-72137). In: SPE Asia Pacific Improved Oil Recovery Conference, Kuala Lumpur, October 6–9.

Petroleum Recovery Institute (PRI), 1995. Prize^{TM} Manual, Version 1.1, June.

Pinto, d.a.C.P.H.L., Silva Jr., M.F., Izetti, R.G., Guimarães, G.B., 2006. Integrated Multizone Low-Cost Intelligent Completion for Mature Fields (SPE-99948). In: SPE Intelligent Energy Conference and Exhibition, Amsterdam, April 11–13.

Pitts, M.J., Dowling, P., Wyatt, K., Surkalo, H., Adams, C., 2006. Alkaline-Surfactant-Polymer Flood of the Tanner Field (SPE-100004). In: SPE/DOE Symposium on Improved Oil Recovery, Tulsa, April 22–26.

Pitts, M.J., Wyatt, K., Surkalo, H., 2004. Alkaline-Polymer Flooding of David Pool, Lloydminster Alberta (SPE-89386). In: SPE/DOE Fourteenth Symposium on IOR, Tulsa, April 17–21.

Platt, J.D., 2008. Minimizing Environmental Impacts in the Arctic: 30 Years of Oil Development on the North Slope of Alaska (SPE-111957). In: SPE International Conference on Health, Safety, and Environment in Oil and Gas Exploration and Production, Nice, April 15–17.

Poellitzer, S., Florian, T., Clemens, T., 2009. Revitalising a Medium Viscous Oil Field by Polymer Injection, Pirawarth Field, Austria (SPE-120991). In: EUROPEC/EAGE Conference and Exhibition, Amsterdam, June 8–11.

Point Carbon, 2009a. EUA OTC Assessment (EUR/t), Point Carbon. August 21. http://www.pointcarbon.com/.

Point Carbon, 2009b. Carbon A–Z—glossary of keywords. Point Carbon. August 21. http://www.pointcarbon.com/1.266906#E.

Poll, P., Zou, J., Chianis, J., 2009. Contemporary Challenges and Solutions for Post-Katrina Gulf of Mexico Spar Design (OTC-20238). In: Offshore Technology Conference, Houston, May 4–7.

Poncet, J., Cholin, F., Nicoletis, S., Durandeau, M., 2002. Enhanced Oil Recovery by Gas Injection in HP/HT Reservoir: Deep Jusepin Field Case (Venezuela). In: 23rd Annual Workshop & Symposium Collaborative Project on EOR, International Energy Agency, Caracas, September 8–12.

Portwood, J.T., 2005. The Kansas Arbuckle Formation: Performance Evaluation and Lessons Learned from More Than 200 Polymer-Gel Water-Shutoff Treatments (SPE-94096). In: SPE Production Operations Symposium, Oklahoma City, April 16–19.

Pratap, M., Gauma, M., 2004. Field Implementation of Alkali-Surfactant-Polymer (ASP) Flooding: A Maiden Effort in India (SPE-88455). In: SPE Asia Pacific Oil and Gas Conference and Exhibition, Perth, October 18–20.

Pratap, M., Roy, R.P., Gupta, R.K., Singh, D., 1997. Field Implementation of Polymer EOR Technique—A Successful Experiment in India (SPE-38872). In: SPE Annual Technical Conference and Exhibition, San Antonio, October 5–8.

PTAC, 2003. CO_2 from Industrial Sources to Commercial Enhanced Oil and Gas Recovery. In: Petroleum Technology Alliance of Canada (PTAC) CO_2 Conference, Calgary, October 1–2.

Pu, H., Xu, Q., 2009. An Update and Perspective on Field-Scale Chemical Floods in Daqing Oilfield, China (SPE-118746). In: SPE Middle East Oil and Gas Show and Conference, Bahrain, March 15–18.

Puckett, D., 2009. BP's Pushing Reservoir Limits Flagship: a Focus on BrightWater. In: ACI Optimising EOR Strategy, London, March 11–12. 2009.

Putnam, P.E., Christensen, S., 2004. Geological Influences on Steam Chamber Growth Within Steam-Assisted Gravity Drainage Reservoirs, Lower Cretaceous McMurray Formation, Northeastern Alberta. In: Annual AAPG-SEPM Convention. (Proc P A114), Dallas, April 18–21.

Putnam, P.E., Christensen, S.L., 1999. McMurray Formation SAGD (Steam-Assisted Gravity Drainage) Reservoirs in Northeastern Alberta: Comparative Architecture and Performance. In: CSPG-Canadian Heavy Oil Association-CWLS Joint Conference (ICE 2004), Calgary, May 31–June 4, 2004.

Qiao, Q., Gu, H., Li, D., Dong, L., 2000. The Pilot Test of ASP (Alkaline/Surfactant/Polymer) Combination Flooding in Karamay Oil Field (SPE-64726). In: SPE Oil & Gas International Conference, Beijing, November 7–10.

Qu, Z., Zhang, Y., Zhang, X., Dai, J., 1998. A Successful ASP (Alkaline/Surfactant/Polymer) Flooding Pilot in Gudong Oil Field (SPE-39613). In: 11th SPE/DOE Improved Oil Recovery Symposium, Tulsa, April 19–22.

Quint, E. Monitoring Contact Movement during Depressurization of the Brent Field (SPE-56951). In: Offshore Europe Oil and Gas Exhibition and Conference Aberdeen, September 7–10.

Railroad Commission of Texas (RRC), 2007. Texas Severance Tax Incentives: Present Incentive Programs. November. http://www.rrc.state.tx.us/programs/og/presenttax.php.

Ramlal, V., 2004. Enhanced Oil Recovery by Steamflooding in a Recent Steamflood Project, Cruse "E" Field, Trinidad (SPE-89411). In: 14th SPE/DOE IOR Symposium, Tulsa, April 17–21.

Ranson, A., Alvarado, V., Manrique, E., 2002. New way for Oil Recovery Methods Screening Using Reservoir Analogs and Machine Learning. Unsolicited Patent, October.

Ranson, A., Hernández, K.Y., Matheus, A., Vivas, A.A., 2001. A New Approach to Identifying Operational Conditions in Multivariable Dynamic Processes Using Multidimensional Projection Techniques (SPE-69523). In: SPE Latin American and Caribbean Petroleum Engineering Conference, Buenos Aires, March 25–28.

Rathman, M.P., McGuire, P.L., Carlson, B.H., 2006. Unconventional EOR Program Increases Recovery in Mature WAG Patterns at Prudhoe Bay (SPE-10042). In: SPE/DOE Symposium on Improved Oil Recovery, Tulsa, April 22–26.

Redman, R.S., 2002. Horizontal Miscible Water Alternating Gas Development of the Alpine Field, Alaska (SPE-76819). In: SPE Western Regional/AAPG Pacific Section Joint Meeting, Anchorage, May 20–22.

Remenyi, I., Szittar, A., Udvardi, G., 1995. CO_2 IOR in the Szank Field Using CO_2 from Sweetening Plant. In: Eighth European Symposium on Improved Oil Recovery, Vienna, May 15.

Rivero, J.A., Mamora, D.D., 2007. Oil Production Gains for Mature Steamflooded Oil Fields Using Propane as a Steam Additive and a Novel Smart Horizontal Producer (SPE-110538). In: SPE Annual Technical Conference and Exhibition, Anaheim, November 11–14.

Rodríguez, F., Christopher, C.A., 2004. Overview of Air Injection Potential for PEMEX (Paper No. 89612). In: AAPG International Conference, Cancun, October 24–27.

Rodríguez, R., Bashbush, J.L., Rincón, A., 2008. Feasibility of Using Electrical Downhole Heaters in Faja Heavy Oil Reservoirs (SPE-117682). In: SPE International Thermal Operations and Heavy Oil Symposium, Calgary, October 20–23.

Rossi, A., 2008. Lanzan "Petróleo Plus" para Subir la Producción. Clarin.com, El Pais, November 12. http://www.servicios.clarin.com/notas/jsp/clarin/v9/notas/imprimir.jsp?pagid=1801230.

Rottenfusser, B., Ranger, M.A., 2004. Geological Comparison of Six Projects in the Athabasca Oil Sands. In: CSPG-Canadian Heavy Oil Association-CWLS Joint Conference. (ICE2004), Calgary, May 31–June 4.

Royce, B., Kaplan, E., Garrell, M., Geffen, T.M., 1984. Enhanced Oil Recovery Water Requirements. Environment Geochemical Health Journal 6 (2), 44–53.

Roychaudhury, S., Rao, N.S., Sinha, S.K., Sur, S., Gupta, K.K., Sapkal, A.V., Jain, A.K., Saluja, J.S., 1997. Extension of In-Situ Combustion Process from Pilot to Semi-Commercial Stage in Heavy Oil Field of Balol (SPE-37547). In: SPE International Thermal Operations and Heavy Oil Symposium, Bakersfield, February 10–12.

Sahin, S., Kalfa, U., Celebioglu, D., 2008. Bati Raman Field Immiscible CO_2 Application— Status Quo and Future Plans (SPE-106575). In: SPE Reservoir Evaluation & Engineering 11 (4), 778–791.

Sahuquet, B.C., Ferrier, J.J., 1982. Steam-Drive Pilot in a Fractured Carbonated Reservoir: Lacq Superieur Field (SPE-9453). Journal of Petroleum Technology 34 (4), 873–880.

Sahuquet, B.C., Spreux, A.M., Corre, B., Guittard, M.P., 1990. Steam Injection in a Low-Permeability Reservoir Through a Horizontal Well in Lacq Superieur Field (SPE-20526). In: SPE Annual Technical Conference and Exhibition, New Orleans, September 23–26.

Samsudin, Y., Darman, N., Husain, D., Hamdan, M.K., 2005. Enhanced Oil Recovery in Malaysia: Making It a Reality (Part II) (SPE-95931). In: SPE International Improved Oil Recovery Conference in Asia Pacific, Kuala Lumpur, December 5–6.

Sánchez, J.L., Astudillo, A., Rodríguez, F., Morales, J., Rodríguez, A., 2005. Nitrogen Injection in the Cantarell Complex: Results After Four Years of Operation (SPE-97385). In: SPE Latin American and Caribbean Petroleum Engineering Conference, Rio de Janeiro, June 20–23.

Sarma, H., Das, S., 2009. Air Injection Potential in Kenmore Oilfield in Eromanga Basin, Australia: A Screening Study Through Thermogravimetric and Calorimetric Analyses (SPE-120595). In: SPE Middle East Oil and Gas Show and Conference, Bahrain, March 15–18.

Saskatchewan Energy and Resources, 2009. Carbon Dioxide EOR in Saskatchewan: Overview, Challenges and Opportunities CO_2 EOR Initiative Information Session. Calgary, March 6. www.er.gov.sk.ca.

Satter, A., Thakur, G., 1994. Integrated Petroleum Reservoir Management: A Team Approach. PennWell.

Satter, A., Iqbal, G.M., Buchwalter, J.L., 2008. Practical Enhanced Reservoir Engineering: Assisted with Simulation Software. PennWell.

Scharf, C., Clemens, T. 2006. CO_2-Sequestration Potential in Austrian Oil and Gas Fields (SPE-100176). In: SPE EUROPEC/EAGE Annual Conference and Exhibition, Vienna, June 12–15.

Schiozer, D.J., Mezzomo, C.C., 2003. Methodology for Field Development Optimization with Water Injection (SPE-82021). In: SPE Hydrocarbon Economics and Evaluation Symposium, Dallas, April 5–8.

Schneider, C., Shi, W.A., Miscible, W.A.G., 2005. Project Using Horizontal Wells in a Mature Offshore Carbonate Middle East Reservoir (SPE-93606). In: SPE Middle East Oil and Gas Show and Conference, Bahrain, March 12–15.

Scott, G.R., 2002. Comparison of CSS and SAGD Performance in the Clearwater Formation at Cold Lake (SPE/CIM/CHOA-79020). In: SPE/PS-CIM/CHOA International Thermal Operations and Heavy Oil Symposium and International Horizontal Well Technology Conference, Calgary, November 4–7.

Seccombe, J.C., Lager, A., Webb, K., Jerauld, G., Fueg, E., 2008. Improving Waterflood Recovery: LoSal^{TM} EOR Field Evaluation (SPE-113480). In: SPE/DOE Symposium on Improved Oil Recovery, Tulsa, April 20–23.

Securities and Exchange Commission (SEC), 2008. Modernization of Oil and Gas Reporting, December. *http://www.sec.gov/rules/final/finalarchive/finalarchive2008.shtml.*

Securities and Exchange Commission (SEC), 2009. Form 8-K: LL&E Royalty Trust (LRT), Washington, DC, filed February 9.

Sedaee, B., Rashidi, F., 2006. Application of the SAGD to an Iranian Carbonate Heavy-Oil Reservoir (SPE-100533). In: SPE Western Regional/AAPG Pacific Section/GSA Cordilleran Section Joint Meeting, Anchorage, May 8–10.

Selamat, S., Teletzke, G.F., Patel, P.D., Darman, N., Suhaimi, M.A., 2008. EOR: The New Frontier in the Malay Basin Development (IPTC-12805). In: International Petroleum Technology Conference, Kuala Lumpur, December 3–5.

Sengupta, T.K., Singh, R., Avtar, R., Singh, K., 2001. Successful Gas Shut Off Using Latest Gel Technology in an Indian Offshore Carbonate Field—A Case Study (SPE-72118). In: SPE Asia Pacific Improved Oil Recovery Conference, Kuala Lumpur, October 6–9.

Senocak, D., Pennell, S.P., Gibson, C.E., Hughes, R.G., 2008. Effective Use of Heterogeneity Measures in the Evaluation of a Mature CO_2 Flood (SPE-113977). In: SPE/DOE Symposium on Improved Oil Recovery, Tulsa, April 20–23.

Shafiei, A., Dusseault, M.B., Memarian, H., Samimi Sahed, B., 2007. Production Technology Selection for Iranian Naturally Fractured Heavy Oil Reservoirs (Paper 2007-145). In: Petroleum Society's Eighth Canadian International Petroleum Conference (58th Annual Technical Meeting), Calgary, June 12–14.

Sharma, S.K., Kak, H.L., Meena, H.L., Pratap, V., 2003. EOR Process In Balol-Santhal Fields, India's Honeymoon With In-Situ Combustion: An Overview. In: Fifth Indian Oil Corporation International Petroleum Conference (Petrotech-2003), New Delhi, January 9–12.

Sharp, J.M., 1975. The Potential of Enhanced Oil Recovery Processes (SPE-5557). In: SPE Fall Meeting of the Society of Petroleum Engineers of AIME, Dallas, September 28–October 1.

Shecaira, F.S., Branco, C.C.M., de Souza, A.L.S., Pinto, A.C.C., de Hollleben, C.R.C., Johann, P.R.S., 2002. IOR: The Brazilian Perspective (SPE-75170). In: SPE/DOE Improved Oil Recovery Symposium, Tulsa, April 13–17.

Shi, W., Corwith, J., Bouchard, A., Bone, R., Reinbold, E., Kuparuk, M., 2008. WAG Project After 20 Years (SPE-113933). In: SPE/DOE Symposium on Improved Oil Recovery, Tulsa, April 20–23.

Shin, H., Polikar, M., 2005. Fast- SAGD Application in the Alberta Oil Sands Areas (Paper 2005-173). In: Petroleum Society's Canadian International Petroleum Conference (56th Annual Technical Meeting), Calgary, June 7–9.

Shokir, E.l.M., Goda, H.M., Sayyouh, M.H., Fattah, K.h., 2002. A Selection and Evaluation EOR Method Using Artificial Intelligence (SPE-79163). In: SPE: Annual International Conference and Exhibition, Abuja, Nigeria, August 5–7.

Shyam, D., Arun, N., Kelvin, R., 1995. Environmental Management Strategies for an Enhanced Oil Recovery Project, Trinidad (SPE-30689). In: SPE Annual Technical Conference and Exhibition, Dallas, October 22–25.

Sierra, R., Tripathy, B., Bridges, J.E., Farouq Ali, S.M., 2001. Promising Progress in Field Application of Reservoir Electrical Heating Methods (SPE-69709). In: SPE International Thermal Operations and Heavy Oil Symposium, Margarita Island, Venezuela, March 12–14.

Singhal, A.K., Ito, Y., Kasraie, M., 1998. Screening and Design Criteria for Steam Assisted Gravity Drainage (SAGD) Projects (SPE-50410). In: SPE International Conference on Horizontal Well Technology, Calgary, November 1–4.

Skinner, D.C., 1999. Introduction to Decision Analysis. Probabilistic Publishing.

Smith, D.D., Giraud, M.J., Kemp, C.C., McBee, M., Taitano, J.A., Winfield, M.S., Portwood, J.T., Everett, D.M., 2006. The Succesful Evolution of Anton Irish Conformance Efforts (SPE-103044). In: SPE Annual Technical Conference and Exhibition, San Antonio, September 24–27.

Snell, J.S., Close, A.D., 1999. Yates Field Steam Pilot Applies Latest Seismic and Logging Monitoring Techniques (SPE-56791). In: SPE Annual Technical Conference, Houston, October 3–6.

Snell, J.S., Wadleigh, E.E., Tilden, J., 2000. Fracture Characterization a Key Factor in Yates Steam Pilot Design and Implementation (SPE-59060). In: SPE International Petroleum Conference and Exhibition, Villahermosa, Mexico, February 1–3.

SPE/WPC/AAPG/SPEE, 2007. Petroleum Resources Management System—PRSM. Oil and Gas Reserves Committee of the Society of Petroleum Engineers.

SPE E & P Glossary. http://www.spe.org/glossary/wiki/doku.php.

Spildo, K., Skauge, A., Aarra, M.G., Tweheyo, M.T., 2009. A New Polymer Application for North Sea Reservoirs (SPE-113460). SPE Reservoir Evaluation & Engineering 12 (3), 427–432.

Stalder, J.L., 2008. Thermal Efficiency and Acceleration Benefits of Cross SAGD (XSAGD)— (SPE-117244). In: SPE International Thermal Operations and Heavy Oil Symposium, Calgary, October 20–23.

Stalkup Jr., F.I., 1983. Status of Miscible Displacement (SPE-9992). Journal of Petroleum Technology 35 (4), 815–826.

Stephenson, D.J., Graham, A.G., Luhning, R.W., 1993. Mobility Control Experience in the Joffre Viking Miscible CO_2 Flood (SPE-23598). SPE Reservoir Engineering 8 (3), 183–188.

Stirpe, M.T., Guzman, J., Manrique, E., Alvarado, V., 2004. Cyclic Water Injection Simulations for Evaluations of Its Potential in Lagocinco Field (SPE-89378). In: SPE/DOE Symposium on Improved Oil Recovery, Tulsa, April 17–21.

Stokka, S., Oesthus, A., Frangeul, J., 2005. Evaluation of Air Injection as an IOR Method for the Giant Ekofisk Chalk Field (SPE-97481). In: SPE International Improved Oil Recovery Conference in Asia Pacific, Kuala Lumpur, December 5–6.

Stripe, J.A., Arisaka, K., Durandeau, M., 1993. Integrated Field Development Planning Using Risk and Decision Analysis to Minimise the Impact of Reservoir and Other Uncertainties: A Case Study (SPE-25529). In: Eighth SPE Middle East Oil Show and Conference, Manama, Bahrain, April 3–6.

Sudirman, S.B., Samsudin, Y.B., Darman, N.H., 2007. Planning for Regional EOR Pilot for Baram Delta, Offshore Sarawak, Malaysia: Case Study, Lesson Learnt and Way Forward (SPE-109220). In: SPE Asia Pacific Oil and Gas Conference and Exhibition, Jakarta, October 30–November 1.

Surguchev, L., Li, L., 2000. IOR Evaluation and Applicability Screening Using Artificial Neural Networks (SPE-59308). In: SPE/DOE Improved Oil Recovery Symposium, Tulsa, April 3–5.

Szabo, J.D., Meyers, K.O., 1993. Prudhoe Bay: Development History and Future Potential (SPE-26053). In: SPE Western Regional Meeting, Anchorage, May 26–28.

Tabary, R., Fornari, A., Bazin, B., Bourbiaux, B., Dalmazzone, C., 2009. Improved Oil Recovery With Chemicals in Fractured Carbonate Formations (SPE-121668). In: SPE International Symposium on Oilfield Chemistry, The Woodlands, Texas, April 20–22.

Taber, J.J., Martin, F.D., 1983. Technical Screening Guide for Enhanced Oil Recovery (SPE-12069). In: 58th Annual SPE Technical Conference, San Francisco, October 5–8.

Taber, J.J., Martin, F.D., Seright, R.S., 1997a. EOR Screening Criteria Revisited—Part I: Introduction to Screening Criteria and Enhanced Oil Recovery Projects (SPE-35385). SPE Reservoir Engineering 12 (3), 189–198.

Taber, J.J., Martin, F.D., Seright, R.S., 1997b. EOR Screening Criteria Revisited—Part II: Applications and Impact of Oil Prices (SPE-39234). SPE Reservoir Engineering 12 (3), 199–205.

Taggart, D.L., Russell, G.C., 1981. Sloss Micellar/Polymer Flood Post Test Evaluation Well (SPE-9781). In: SPE/DOE Enhanced Oil Recovery Symposium, Tulsa, April 5–8.

Talukdar, S., Instefjord, R., 2008. Reservoir Management of the Gullfaks Main Field (SPE-113260). In: EUROPEC/EAGE Conference and Exhibition, Rome, June 9–12.

Tealdi, L., Obondoko, G., Isella, D., Baldini, D., Baioni, A., Okassa, F., Pace, G., Itoua-Konga, F., Rampoldi, M., 2008. The Kitina Mature Offshore Field Rejuvenation: Massive Multistage Hydraulic Fracturing, Long Reach Wells, Improved Oil Recovery Techniques (SPE-113609). In: EUROPEC/EAGE Conference and Exhibition, Rome, June 9–12.

Teletzke, G.F., Wattenbarger, R.C., Wilkinson, J.R., 2008. Enhanced Oil Recovery Pilot Testing Best Practices (SPE-118055). In: SPE International Petroleum Exhibition and Conference, Abu Dhabi, November 3–6.

Teramoto, T., Takabayashi, K., Onishi, T., Okatsu, K., 2005. Air Injection EOR in Highly Water Saturated Oil Reservoir. In: IEA Collaborative Project on Enhanced Oil Recovery 25th Annual Workshop and Symposium, Norway, September 5–8.

Terzian, G.A., Enright, J.M., Brashear, J.P., 1995. Financial Incentives for Marginal Oil and Gas Production (SPE-30045). In: SPE Hydrocarbon Economics and Evaluation Symposium, Dallas, March 26–28.

Thakur, G.C., Satter, A., 1998. Integrated Waterflood Asset Management. PennWell.

Thomas, S., 2008. Enhanced Oil Recovery—An Onverview. Oil & Gas Science and Technology—Rev. IFP 63 (1), 9–19.

Thompson, R.S., Wright, J.D., 1984. Oil Property Evaluation, second ed. Thompson-Wright Associates.

Tiwari, D., Marathe, R.V., Patel, N.K., Ramachandran, K.P., Maurya, C.R., Tewari, P.K., 2008. Performance of Polymer Flood in Sanand Field, India—A Case Study (SPE-114878). In: SPE Asia Pacific Oil and Gas Conference and Exhibition, Perth, October 20–22.

Tovar, J.J., Zerpa, L., Guerra, E., 1999. Impact of Formation Damage on Sand Production in Deep Eocene Reservoirs, Lake Maracaibo, Venezuela: A Case History (SPE-54757). In: SPE European Formation Damage Conference, The Hague, May 31–June 1.

Trantham, J.C., 1983. Prospects of Commercialization, Surfactant/Polymer Flooding, North Burbank Unit, Osage County, OK (SPE-9816). Journal of Petroleum Technology 35 (5), 872–880.

Trantham, J.C., Patterson, H.L., Boneau, D.F., 1978. The North Burbank Unit, Tract 97 Surfactant/Polymer Pilot Operation and Control (SPE-6746). Journal of Petroleum Technology 30 (7), 1068–1074.

Tyler, N., Finley, R.J., 1991. Architectural Controls on the Recovery of Hydrocarbons from Sandstone Reservoirs. SEPM Concepts in Sedimentology and Paleontology 3, 3–7.

U.S. Department of Energy, 1995. Toris: Data Preparation Guidelines. U. S. DoE Report NIPER/BDM-0042, March.

Vanegas, J.W., Clayton, P., Deutsch, V., Cunha, L.B., 2009. Transference of Reservoir Uncertainty in Multi SAGD Well Pairs (SPE-124153). In: SPE Annual Technical Conference and Exhibition, New Orleans, October 4–7.

Vargo, J., Turner, J., Vergnani, B., Pitts, M.J., Wyatt, K., Surkalo, H., Patterson, D., 2000. Alkaline-Surfactant-Polymer Flooding of the Cambridge Minnelusa Field (SPE-68285). SPE Reservoir Evaluation & Engineering 3 (6), 552–558.

Vassilellis, G.D., 2009. Roadmap to Monetization of Unconventional Resources (SPE-121968). In: EUROPEC/EAGE Conference and Exhibition, Amsterdam, June 8–11.

Veil, J.A., Quinn, J.J., Garcia, J.P., 2009. Water Issues Relating to Heavy Oil Production (SPE-120630). In: SPE Americas E&P Environmental and Safety Conference, San Antonio, March 23–25.

Velasquez, D., Rey, O., Manrique, E., 2006. An Overview of Carbon Dioxide Sequestration in Depleted Oil and Gas Reservoirs in Florida, USGS Petroleum Province 50. In: Fourth LACCEI International Latin American and Caribbean Conference for Engineering and Technology, Mayagüez, Puerto Rico, June 21–23.

Voneiff, G.W., Cipolla, C., 1996. A New Approach to Large-Scale Infill Evaluations Applied to the OZONA (Canyon) Gas (SPE-35203). In: SPE Permian Basin Oil and Gas Recovery Conference, Midland, March 27–29, 1996.

Waheed, A., El-Assal, H., Negm, E., Sanad, M., Sanad, O., Tuchscherer, G., Sayed, M., 2001. Practical Methods to Optimizing Production in a Heavy-Oil Carbonate Reservoir: Case Study From Issaran Field, Eastern Desert, Egypt (SPE-69730). In: SPE International Thermal Operations and Heavy Oil Symposium, Margarita Island, Venezuela, March 12–14.

Wang, D., Cheng, J., Wu, J., Wang, G., 2002. Experiences Learned after Production of more than 300 million Barrels of Oil by Polymer Flooding in Daqing Oil Field (SPE-77693). In: SPE Annual Technical Conference and Exhibition, San Antonio, September 29–October 2.

Wang, D., Huanzhong Dong, H., Changsen, L.v., Fu, X., Nie, J., 2009b. Review of Practical Experience by Polymer Flooding at Daqing (SPE-114342). SPE Reservoir Evaluation & Engineering 12 (3), 470–476.

Wang, D., Zhang, Y., Yongjian, L., Hao, C., Guo, M., 2009a. The Application of Surfactin Biosurfactant as Surfactant Coupler in ASP Flooding in Daqing Oil Field (SPE-119666). In: SPE Middle East Oil and Gas Show and Conference, Bahrain, March 15–18, 2009.

Watson Jr., C.C., Johnson, M.E., 2006. Real-Time Prediction of Shut-In Production from Hurricanes in the Gulf of Mexico (SPE-102374). In: SPE Annual Technical Conference and Exhibition, San Antonio, September 24–27.

Watts, B.C., Hall, T.F., Petri, D.J., 1997. The Horse Creek Air Injection Project: An Overview (SPE-38359). In: SPE Rocky Mountain Regional Meeting, Casper, May 18–21.

Webb, K.J., Black, C.J.J., Tjetland, G., 2005. A Laboratory Study Investigating Methods for Improving Oil Recovery in Carbonates (IPTC-10506). In: International Petroleum Technology Conference, Doha, Qatar, November 21–23.

Webb, C., Nagghappan, L.N.S.P., Smart, G., Hoblitzell, J., Franks, R., 2009. Desalination of Oilfield-Produced Water at the San Ardo Water Reclamation Facility, CA (SPE-121520). In: SPE Western Regional Meeting, San Jose, March 24–26.

Weiss, W.W., Balch, R.S., Stubbs, B.A., 2000. How Artificial Intelligence Methods Can Forecast Oil Production (SPE-75143). In: SPE Improved Oil Recovery Symposium, Tulsa, April 13–17.

Welsh, M.B., Begg, S.H., 2007. Modeling the Economic Impact of Cognitive Biases on Oil and Gas Decisions (SPE-110765). In: SPE Annual Technical Conference and Exhibition, Anaheim, November 11–14.

Welsh, M.B., Bratvold, R.B., Begg, S.H., 2005. Cognitive Biases in the Petroleum Industry: Impact and Remediation (SPE-96423). In: SPE Annual Technical Conference and Exhibition, Dallas, October 9–12.

WHITESANDS Project—Petrobank Energy and Resources Ltd., 2009 (April 16). *http://www. petrobank.com/hea-whitesandsproject.html*.

Wo, S., Yin, P., Blakeney-DeJarnett, B., Mullen, C.E., 2008. Simulation Evaluation of Gravity Stable CO_2 Flooding in the Muddy Reservoir at Grieve Field, Wyoming (SPE-113482). In: SPE/DOE Symposium on Improved Oil Recovery Tulsa, April 20–23.

Wo, S., Whitman, L.D., Steidtmann, J.R., 2009. Estimates of Potential CO_2 Demand for CO_2 EOR in Wyoming Basins (SPE-122921). In: SPE Rocky Mountain Petroleum Technology Conference, Denver, April 14–16.

Wozinak, D.A., Wing, J.L., Schrider, L.A., 1997. Infill Reserve Growth Resulting from Gas Huff-n-Puff and Infill Drilling—A Case History (SPE-39214). In: SPE Eastern Regional Meeting, Lexington, Kentucky, October 22–24.

Wyoming Geological Association, 2000. Publications on DVD, 1946–2000. AAPG/Datapages, Tulsa.

Wyoming Oil and Gas Conservation Commission, 2008a. 051208 Well Header database. Retrieved May 12, 2008, from the Wyoming Oil and Gas Conservation Commission Web site. *http://wogcc.state.wy.us/urecordsMenu.cfm?Skip='Y'&oops=49*.

Wyoming Oil and Gas Conservation Commission, 2008b. Field Production Data. Multiple retrievals 2007 through 2008, from the Wyoming Oil and Gas Conservation Commission Web site. *http://wogcc.state.wy.us/FieldMenu2.cfm*.

Xia, T.X., Greaves, M., Turta, A.T., 2002. Injection Well—Producer Well Combinations in THAI 'Toe-to-Heel' Air Injection (SPE-75137). In: SPE/DOE Improved Oil Recovery Symposium Tulsa, April 13–17.

Xia, T.X., Greaves, M., Werfilli, W.S., Rathbone, R.R., 2002. Downhole Conversion of Lloydminster Heavy Oil Using THAI-CAPRI Process (SPE-78998). In: SPE International Thermal Operations and Heavy Oil Symposium and International Horizontal Well Technology Conference, Calgary, November 4–7.

Xiang, W., Zhou, W., Zhang, J., Yang, G., Jiang, W., Sun, L., Li, J., 2008. The Potential of CO_2-EOR in China Offshore Oilfield (SPE-115060). In: SPE Asia Pacific Oil and Gas Conference and Exhibition, Perth, October 20–22.

Xie, X., Economides, M.J., 2009. The Impact of Carbon Geological Sequestration (SPE-120333). In: SPE Americas E&P Environmental and Safety Conference, San Antonio, March 23–25.

x-rates.com, 2009. Exchange Rates. August 21. *http://www.x-rates.com/*.

Yang, H.D., Wadleigh, E.E., 2000. Dilute Surfactant IOR-Design Improvement for Massive, Fractured Carbonate Applications (SPE-59009). In: SPE International Petroleum Conference and Exhibition, Villahermosa, Mexico, February 1–3.

Zainal, S., Manap, A.A.A., Hamid, P.A., Othman, M., Chong, M.O., Yahya, A.W., Darman, N., Sai, R.M., Offshore Chemical, E.O.R., 2008. The Role of an Innovative Laboratory Program in Managing Result Uncertainty to Ensure the Success of a Pilot Field Implementation (SPE-113154). In: EUROPEC/EAGE Conference and Exhibition, Rome, June 9–12.

Zang, X., Guan, W., Li, X., Liu, Z., Meng, J., Jiang, D., 2009. Novel Injecting Concentration Design Method For Polymer Flooding in Heterogeneous Reservoirs (SPE-123404). In: SPE Asia Pacific Oil and Gas Conference & Exhibition, Jakarta, August 4–6.

Zeidouni, M., Moore, M., Keith, D., 2009. Guidelines for a Regulatory Framework to Accommodate Geological Storage of CO_2 in Alberta (SPE-121000). In: SPE Americas E&P Environmental and Safety Conference, San Antonio, March 23–25.

Zhang, J., Wang, K., He, F., Zhang, F., 1999. Ultimate Evaluation of the Alkali/Polymer Combination Flooding Pilot Test in Xing Long Tai Oil Field (SPE-57291). In: SPE Asia Pacific IOR Conference, Kuala Lumpur, October 25–26.

Zhdanov, S.A., Amiyan, A.V., Surguchev, L.M., Castanier, L.M., Hanssen, J.E., 1996. Application of Foam for Gas and Water Shut-Off: Review of Field Experience (SPE-36914). In: SPE European Petroleum Conference, Milan, October, 22–24.

Ziegler, V.M.A., 1987. Comparison of Steamflood Strategies: Five-Spot Pattern vs. Inverted Nine-Spot Pattern (SPE-13620). SPE Reservoir Engineering 2 (4), 549–558.

Zhijian, Q., Yigen, Z., Xiansong, Z., Jialin, D., 1998. A Successful ASP Flooding Pilot in Gudong Oil Field (SPE-39613). In: SPE/DOE IOR Symposium, Tulsa, April 19–22.

Zhou, Y., Lin, Y., Chen, G., 1998. Steam Stimulation Pilot in the Cao-20 Fracture Limestone with Extra Heavy Oil, Paper 1998-131. In: Seventh UNITAR International Conference on Heavy Crude and Tar Sands, Beijing, October 27–30.

Zubari, H.K., Sivakumar, V.C.B., 2003. Single Well Tests to Determine the Efficiency of Alkaline-Surfactant Injection in a Highly Oil-Wet Limestone Reservoir (SPE-81464). In: SPE 13th Middle East Oil Show & Conference, Bahrain, April 5–8.

Index

Note: Page numbers followed by *f* and *t* indicate figures and tables, respectively.

Printed and bound by CPI Group (UK) Ltd, Croydon, CR0 4YY

03/10/2024

01040434-0013